Resource Utilization of Spent Potlining and Anode Dust in Primary Aluminum Industry

铝电解工业废槽衬与阳极炭渣的资源化利用

Xie Mingzhuang　　Zhao Hongliang　　Liu Fengqin
谢明壮　　　　　赵洪亮　　　　　刘风琴　著

本书数字资源

Beijing
Metallurgical Industry Press
2024

Metallurgical Industry Press
39 Songzhuyuan North Alley, Dongcheng District, Beijing 100009, China

Copyright © Metallurgical Industry Press 2024. All rights reserved.

No part of this publication may be reproduced or transmitted in any form or by any means, electronic or mechanical, including photocopying, recording, or any information storage and retrieval system, without permission in writing from the copyright owner.

图书在版编目(CIP)数据

铝电解工业废槽衬与阳极炭渣的资源化利用 = Resource Utilization of Spent Potlining and Anode Dust in Primary Aluminum Industry: 英文/谢明壮, 赵洪亮, 刘风琴著. —北京: 冶金工业出版社, 2024.1
ISBN 978-7-5024-9716-3

Ⅰ.①铝… Ⅱ.①谢… ②赵… ③刘… Ⅲ.①氧化铝电解—危险废弃物—废物综合利用—英文 Ⅳ.①X758

中国国家版本馆 CIP 数据核字(2024)第 034924 号

Resource Utilization of Spent Potlining and Anode Dust in Primary Aluminum Industry

出版发行	冶金工业出版社	
地　　址	北京市东城区嵩祝院北巷 39 号	电　话　(010) 64027926
网　　址	www.mip1953.com	邮　编　100009
		电子信箱　service@ mip1953.com

责任编辑　王　双　　美术编辑　燕展疆　　版式设计　郑小利
责任校对　葛新霞　　责任印制　禹　蕊

北京建宏印刷有限公司印刷
2024 年 1 月第 1 版, 2024 年 1 月第 1 次印刷
710mm×1000mm 1/16; 14.75 印张; 324 千字; 223 页
定价 99.00 元

投稿电话　(010)64027932　投稿信箱　tougao@cnmip.com.cn
营销中心电话　(010) 64044283
冶金工业出版社天猫旗舰店　yjgycbs.tmall.com

(本书如有印装质量问题, 本社营销中心负责退换)

Preface

Aluminum is one of the most productive and versatile non-ferrous metals in the world. It is a highly correlated and indispensable basic raw material in the economic development of a country, widely used in various sectors such as aerospace, transportation, civil construction, packaging, power, electronics, etc. It is a green strategic metal with excellent performance and can be efficiently recycled, which plays an irreplaceable role in the national economy and national defense construction. The primary aluminum is produced using a cryolite-molten salt reduction method. The primary aluminum industry has developed rapidly in recent decades. The rapid economic development has driven strong demand for primary aluminum, which has effectively promoted the rapid development of China's aluminum industry. China has developed into the world's largest producer of alumina, primary aluminum, and carbon industries for aluminum. The production of alumina and carbon for aluminum in China in 2022 is 81.86 Mt and 21.64 Mt, respectively. Data from *World Aluminum* shows that about 40.21 Mt of primary aluminum was produced in China in 2022, which accounted for more than 59% of the world's primary aluminum production.

The huge production scale of the aluminum industry has led to increasingly prominent issues such as resources, energy, and environment. Especially the production of solid waste materials such as spent potlining, anode dust, emitted by China's aluminum industry, is enormous and increasing year by year, causing enormous pressure and impact on ecology and the environment. In 2022, about 365 kt of carbon dust and 900 kt of spent potlining were produced in China. The spent potlining consists of three parts: spent cathode carbon block, waste silicon carbide side block and waste refractory material. Soluble fluorine and cyanide in these substances can be dissolved into the soil and groundwater when they are stacked outdoors and in contact with humid air or rain. It will cause great harm to the local environment and society. At present, these aluminum reduction industrial wastes have been listed in the National Hazardous Waste List in China. The huge amount of industrial hazardous waste is not only a great burden on the environment, but also a serious waste of valuable resources. Prof. Fengqin Liu from University of Science and Technology Beijing has led the *Green and Efficient Light Metal Smelting Team* to conduct extensive and in-depth research on hazardous waste in the aluminum smelting industry for many years. Multiple green and economic process technologies have been proposed for the treatment of different types of solid hazardous waste, and have been applied in several well-known enterprises in China.

This book has 4 chapters. Dr. Mingzhuang Xie, A. Prof. Hongliang Zhao and Prof. Fengqin Liu from University of Science and Technology Beijing are responsible for writing, and overall revision of this book. This book introduces the current situation and technological progress of China's aluminum industry, with a focus on the latest research results of green and low-carbon disposal technologies for two types of hazardous solid waste emitted by the aluminum industry: spent potlining and anode dust. The classification of spent potlining, pyrometallurgical disposal, hydrometallurgical disposal, and thermal behavior analysis of the disposal process were introduced in detail. The anode dust mainly introduces the calcification roasting process to achieve efficient extraction of valuable elements in electrolytes. The roasting process and leaching process with additives have been proposed.

This study was supported by the National Key Research and Development Program of China (2019YFC1908403). The authors express their gratitude to the people who provided assistance during the writing process of this book. The authors express their gratitude to State Power Investment Group Ningxia Energy Aluminum Co., Ltd., Zhejiang Yutao Environmental Protection Technology Co., Ltd., and Inner Mongolia Risheng Renewable Resources Co., Ltd. for their assistance and support in promoting the application of technological achievements. The authors thank Dr. Rongbin Li, Dr. Han Lv, Dr. Tingting Lu, Dr. Zhengping Zuo, Zegang Wu, Guoqing Yu, Jingjing Zhong, Xin Yang, Aijie Li and Shuang Hong from the *Green and Efficient Light Metal Smelting Team* at University of Science and Technology Beijing for participating in the editing work of this book.

Due to limited professional knowledge and the preparation of time, there may be some shortcomings in the book, sincerely hope that readers can correct.

Authors
University of Science and Technology Beijing
December 2023

Contents

Chapter 1 Development and Prospects of China's Aluminum Industry ········· 1

1.1 Current situation of alumina industry production in China ············· 1
   1.1.1 The production and capacity of China's alumina industry ··············· 1
   1.1.2 Current status of alumina production technology ························ 3

1.2 Current situation of aluminum reduction industry in China ············ 6
   1.2.1 The production and capacity of China's aluminum reduction industry ······ 7
   1.2.2 Current status of primary aluminum production technology in China ······ 9

1.3 Solid/Hazardous waste disposal in China's aluminum smelting industry ················ 13
   1.3.1 Current situation of red mud discharge and treatment ················ 13
   1.3.2 Flue gas and wastewater discharge and treatment ···················· 16
   1.3.3 Discharge and treatment of anode carbon residue in aluminum electrolysis process ························ 18
   1.3.4 Emission and treatment of hazardous solid waste ···················· 19

1.4 Sustainable development strategies for China's aluminum smelting industry ················ 22
   1.4.1 Sustainable development strategies for China's alumina industry ······ 22
   1.4.2 Sustainable development strategies for China's aluminium reduction industry ······················ 24

References ·················· 26

Chapter 2 Spent Potlining Disposal Technology ··············· 30

2.1 Overview ················· 30
   2.1.1 Cement industry ··············· 31
   2.1.2 Steel industry ················· 31
   2.1.3 Mineral wool industry ············ 32
   2.1.4 Electric power industry ··········· 32
   2.1.5 Landfill ······················ 33

2.2 Classification, composition, and phase ... 35
2.3 Characterization analysis ... 36
  2.3.1 Composition of spent cathode carbon block ... 36
  2.3.2 Composition of waste silicon carbide side block ... 38
  2.3.3 Composition of waste refractory material ... 39
2.4 Toxic substance footprint ... 40
2.5 Pyrometallurgy treatment technology ... 51
  2.5.1 Experimental study on fluorine separation ... 51
  2.5.2 Kinetics of fluorine separation ... 68
  2.5.3 Thermal conductivity simulation of waste cathode carbon block ... 79
  2.5.4 Numerical simulation of the electro-thermal coupling treatment process ... 92
  2.5.5 Reduction of converter slag with waste cathode carbon block ... 102
  2.5.6 Recycling of spent potlining by different high temperature treatments ... 125
2.6 Hydrometallurgy disposal technology ... 142
  2.6.1 Leaching of waste cathode carbon block ... 142
  2.6.2 Leaching of waste side blocks ... 153
  2.6.3 Leaching of waste refractory materials ... 162
2.7 Analysis on thermal behavior of fluorides and cyanides by TG/DSC-MS & ECSA ... 170
  2.7.1 Materials and characterization ... 173
  2.7.2 TG/DSC-MS measurements and analysis by ECSA ... 175
  2.7.3 Thermal behavior of fluorides under Ar and Ar$-O_2$ atmospheres ... 176
  2.7.4 Thermal behavior of cyanides under Ar and combustion atmospheres ... 179
References ... 184

Chapter 3 Anode carbon residue/electrolyte disposal technology ... 192

3.1 Composition and phase ... 193
3.2 Experiment on calcification and roasting of waste aluminum electrolyte ... 193
  3.2.1 Effect of calcium oxide addition on aluminum and sodium recovery rates ... 193

  3.2.2 Effect of calcination temperature on the recovery rate of aluminum and sodium ……………………………………………………… 202
  3.2.3 Effect of sodium carbonate addition on aluminum and sodium recovery rates …………………………………………………… 207
 3.3 Dissolution experiment of calcination products ……………………… 210
  3.3.1 Effect of dissolution temperature ……………………………………… 211
  3.3.2 Effect of dissolution time ……………………………………………… 212
  3.3.3 Effect of alkali concentration …………………………………………… 213
  3.3.4 Analysis of microscopic morphology of dissolved products ………… 214
 References ………………………………………………………………………… 217

Chapter 4 Outlook …………………………………………………………… 220
 4.1 Resource and energy development strategy of China's aluminum industry ……………………………………………………… 221
 4.2 Strategy for improving core competitiveness of China's aluminum industry ……………………………………………………… 221
 4.3 Environmental development strategy of China's aluminum industry ……………………………………………………… 223

Chapter 1   Development and Prospects of China's Aluminum Industry

1.1 Current situation of alumina industry production in China

The first alumina plant in China was Shandong Aluminum Plant, which adopted sintering technology and was put into operation in 1954. From the 1960s to the 1990s, Zhengzhou Aluminum Plant, Guizhou Aluminum Plant, and Shanxi Aluminum Plant were successively built, all using the Bayer sintering hybrid method developed in China. The first Bayer process alumina plant in China was Pingguo Aluminum Plant, which was put into operation in 1995. However, in the last century, the total annual production of alumina in China was less than 5 Mt.

1.1.1 The production and capacity of China's alumina industry

Since 2001, China's alumina industry has entered a rapid development track, as shown in Fig. 1-1. During the decade from 2000 to 2022, the annual production increased by about 9-fold. In 2022, the annual alumina production reached 81.86 Mt, accounting for 55%–60% of the world's total alumina production[1].

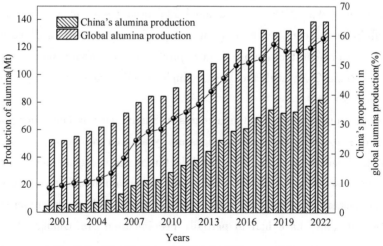

Fig. 1-1 Alumina production in China over the years

In the past decade, China's alumina production has continued to increase rapidly, and production capacity has also increased accordingly. In 2022, it exceeded 80 Mt, and the utilization rate of production capacity has always hovered between 70%–90%.

The rapid growth of alumina production requires the consumption of more and more bauxite resources[2]. China's bauxite reserves are about 1 billion tons, ranking 7th in the world. Bauxite resources are not abundant, and China's bauxite is mainly composed of difficult to treat diaspore, with a low grade.

After decades of mining and utilization, high-quality bauxite with an aluminum silicon ratio of over 7 in the northern region has been basically depleted. At present, the domestic bauxite reserves are relatively small, the grade is poor, the composition is complex, and the processing difficulty is high[3]. This is an important reason for the lack of competitiveness of China's alumina industry, and it is also a serious challenge for the sustainable development of China's alumina industry.

The distribution of alumina production capacity in China has a clear regional concentration characteristic guided by resources. Affected by the distribution of bauxite resources and transportation costs, China's alumina production capacity is mainly distributed in provinces and regions such as Shanxi, Henan, Guangxi, and Guizhou, where bauxite reserves are relatively abundant[4]. Although local bauxite has been basically mined in Shandong Province, due to its coastal advantages, bauxite imports are relatively convenient, and the proportion of alumina production capacity is also large.

Statistical data shows that in 2022, the alumina production of Shandong, Shanxi, Henan, Guangxi, and Guizhou accounted for 34.0%, 24.9%, 12.4%, 15.6% and 6.2% of the national production capacity, respectively. The five provinces collectively accounted for more than 94% of the national total production capacity. In recent years, the regional/industrial concentration of alumina production capacity in China has rapidly increased[5]. At present, the average annual production of alumina plants in China has increased to around 2 Mt, with the largest single alumina plant reaching a world-class scale of over 6 Mt.

However, the construction of alumina plants in Shanxi, Henan, and other places is too intensive and large-scale. The consequences are: years of continuous large-scale mining and use of local bauxite, resulting in a rapid decline in bauxite grade and rapid depletion of high-quality resources. At present, Shanxi and Henan alumina factories have generally suffered from a shortage of bauxite supply and poor grade (with an aluminum silicon ratio of only about 4), and have had to partially use imported ore. Guangxi has abundant bauxite reserves, high grade, and convenient mining, which is conducive to the development of the alumina industry. However, several new alumina plants have been built recently. If local alumina production scale is not controlled, there may also be a shortage of local bauxite supply in the future.

The production of alumina from imported bauxite has become a development trend in China's alumina industry. Almost all alumina production in Shandong Province relies on imported bauxite, and alumina factories in Shanxi, Henan, and other regions are gradually adopting or partially using imported ore to replace domestic ore in the production of alumina.

However, due to the high cost of domestic land transportation for imported mines, the cost lacks competitiveness[6]. In the long run, with the increasing depletion and depletion of domestic bauxite resources, major domestic alumina enterprises have chosen to build alumina plants in ports or overseas to replace inland alumina production capacity. For example, Chalco has basically built an alumina plant with a production capacity of 2 Mt in Fangchenggang, Guangxi, and Weiqiao Aluminum has built an alumina plant with a production capacity of 1 Mt in Indonesia.

The intensive construction of alumina plants in the mainland of China has also brought about another major problem, namely environmental issues, which is the excessive discharge of red mud[7]. The discharge of red mud requires occupying a large area of land. If the alumina plant is too dense and the discharge of red mud is locally too large, it will cause serious pollution[8]. In addition, some areas belong to water shortage areas, and the construction of alumina plants is also severely constrained by the lack of water resources.

In short, the intensive construction and excessive production capacity of alumina plants in some northern regions of China have caused incalculable pressure on resources, energy, and the environment. The use of imported bauxite has become a development trend in China's alumina industry.

1.1.2 Current status of alumina production technology

There are three types of process technologies for alumina production in China: Bayer process, sintering process, and combined process (mixed process and series process)[9]. At present, the Bayer method accounts for 97% of the production of alumina, becoming the main technology in China's alumina industry[10]. The production capacity of the combined method only accounts for 2%, while the sintering method is mainly used for the production of high-value chemical alumina (fine alumina) due to its high energy consumption and operating costs, with a production capacity of only 1%.

Before 1994, the main production methods for alumina in China were the combined method and sintering method. By 2017, the Bayer process had been transformed into the main process, and most of it was used to treat China's insoluble diaspore ore[11]. This transformation was achieved through a large number of persistent technological breakthroughs by Chinese alumina technology workers. The Bayer process is simple, energy-saving, and has high product quality, which has significantly optimized the technical indicators of alumina production in China[12].

Based on the development of Bayer process for treating monohydrate bauxite, China has successfully developed alumina production technologies such as beneficiation Bayer process and lime Bayer process, which are suitable for treating medium to low grade monohydrate bauxite, in response to the increasingly declining grade of bauxite in China. The high-energy sintering process has been gradually phased out.[13].

Over the past decade, with the continuous decline in bauxite grade, energy shortage, and rising prices, significant energy conservation has become the main focus of new technology development for alumina production in China[14]. A series of technologies used to treat low-grade diaspore ore have achieved significant energy-saving goals.[15].

To achieve this goal, China has independently developed Bayer indirect heating and enhanced dissolution technologies for treating diaspore ore at high temperatures (above 260 ℃) and high alkali concentrations (above 230 g/L), including pipeline preheating, retention tank dissolution, full pipeline preheating and dissolution, high-temperature dual flow dissolution, etc., high-efficiency separation technology for red mud using high bond deep cone settling tanks, six effect countercurrent falling film evaporation technology Fluidized roasting (including gas suspension roasting furnace and circulating fluidized roasting furnace) and other technologies, thus opening up the entire process of high-temperature Bayer method for treating diaspore ore and achieving energy conservation and consumption reduction in each main process[16]. In addition, China has independently developed high-efficiency enhanced Bayer process technology to achieve high yield and high cycle efficiency in key processes through the enhancement of process material flow, and to narrow the energy consumption gap with foreign countries through systematic energy-saving methods[17].

The research and application of these energy-saving new technologies, processes, and equipment have continuously reduced the energy consumption of alumina production in China[18]. In 2021, the average energy consumption for producing tons of alumina in China reached 10.06 GJ, of which 2/3 of the production was produced by processing insoluble and low-grade diaspore ore. This is very close to the world average energy consumption of 10.5 GJ.

A major improvement technology completed by China's alumina industry in the past decade is the production of sandy alumina using high concentration seed fractions[19]. Given that China's alumina industry mainly processes diaspore ore, the main conditions for seed decomposition are high alkali concentration, low supersaturation, and a one-stage decomposition process[20]. China has independently developed a low decomposition initial temperature and multi-level intermediate cooling temperature system, a high crystal seed addition amount, and a long decomposition time seed separation process, achieving high yield under low seed supersaturation[21]. At the same time, a decomposition system of particle size prediction and advance regulation by detecting the change rate of the number of grains of fine microcrystalline seeds has been developed to ensure the strict control of alumina particle size and successfully achieve the goal of producing sandy alumina from diaspore ore[22].

In terms of environmental governance, China has innovatively developed red mud dry pile technology. Red mud dry pile refers to the method of using a filter press to filter the red mud discharged into the yard, and then transporting filter cakes with a moisture content of about 30% to the yard for storage using belts or cars[23]. This technology not only recycles more caustic soda from red mud, but also is very beneficial for protecting red mud reservoirs and reducing accidents caused by dam breaks due to excessive water content in red mud. Additionally, the utilization rate of storage capacity in the storage yard is increased, making it easy for reclamation after closure.

Alumina plants in Shandong province, China, as well as some alumina plants in Shanxi, Henan, and Chongqing, use imported bauxite to produce alumina. Imported ore generally has a higher grade (aluminum silicon ratio), but a lower alumina content, mainly consisting of gibbsite type ore and gibbsite boehmite mixed ore[24]. According to the type and phase composition of imported ore, two different Bayer process technologies, low-temperature Bayer process and high-temperature Bayer process, are used for treatment[25-27].

The low-temperature Bayer method is used to treat gibbsite ore, with a dissolution temperature of 140–145 ℃, a retention time of about 1 h, and an alkaline solution concentration of Na_2O_K below 180 g/L. The amount of seed added for seed decomposition is relatively small, and the decomposition time is also short (30–40 h). The product is coarse-grained sandy alumina. The main problem with the low-temperature Bayer process is that it is difficult to recover alumina from aluminum goethite and other aluminum containing minerals, resulting in low alumina recovery rate and high ore consumption[28].

The high-temperature Bayer method mainly deals with the mixed ore of trihydrate aluminite and monohydrate aluminite, with a dissolution temperature of over 240 ℃ and an alkaline solution concentration of over 200 g/L. Seed decomposition generally adopts a one-step method, with a low initial temperature (58-65 ℃), a large amount of seed addition (500-800 g/L), a longer decomposition time, and a higher decomposition yield[29-32]. The main problem with the high-temperature Bayer process is that some non active silicon is converted into active silicon, thereby increasing alkali consumption.

Given that imported ore may become an important raw material for future alumina production in China, it is necessary to conduct in-depth research on the leaching behavior and performance of imported ore, and propose new process equipment suitable for China's production conditions[33].

1.2 Current situation of aluminum reduction industry in China

Fushun Aluminum Plant was the first aluminum electrolysis plant built in New China to produce raw aluminum. It was put into operation in 1954 and used the self baking aluminum electrolysis cell technology that was popular in the world at that time[34]. Until the early 1980s, China successively built eight major aluminum electrolysis plants, including Baotou Aluminum Plant, Jiaozuo Aluminum Plant, Lanzhou Aluminum Plant, Qingtongxia Aluminum Plant, Liancheng Aluminum Plant, and Guizhou Aluminum Plant, all of which used 60 kA capacity self baking aluminum electrolysis cell technology[35]. In 1982, Guizhou Aluminum Plant introduced 160 kA pre baked anode aluminum electrolysis cell technology from Japan and further developed 180 kA pre baked aluminum electrolysis cell technology. Since 1988, with the support of the National Science and Technology Plan, China National Nonferrous Metals Industry Corporation has organized a joint research project-the 280 kA large pre baked aluminum electrolysis cell technology-by Zhengzhou Light Metals Research Institute, Shenyang Aluminum Magnesium Design and Research Institute, and Guiyang Aluminum Magnesium Design and Research Institute. After eight years of continuous research, it was successful in 1996. This technology has become a milestone in the large-scale application of large-scale pre baked aluminum electrolysis cell technology in China. Subsequently, a series of technologies for 300 kA, 400 kA, and 500 kA aluminum electrolytic cells were developed, and even the industrialization of the 600 kA large-scale aluminum electrolytic cell series was achieved for the first time in the world[36]. At the same time, the selfbaking aluminum electrolysis cell technology, which is technologically outdated and heavily polluted, has also been gradually phased out[37].

1.2.1 The production and capacity of China's aluminum reduction industry

The promotion and application of large-scale pre baked aluminum electrolysis cell technology has completely changed the backward situation of China's aluminum electrolysis industry technology, and effectively promoted a significant increase in aluminum electrolysis production[38]. As shown in Fig. 1-2, China's primary aluminum production has accelerated since the 1990s. In the past two decades, private economy has entered the aluminum electrolysis industry on a large scale, leading to a rapid development of China's primary aluminum production. In the more than ten years from 2005 to 2022, China's primary aluminum production increased by about five times.

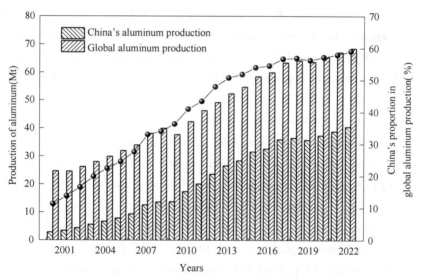

Fig. 1-2 Changes in annual production of primary aluminum in China

Since 2001, China has always maintained its position as the world's largest producer of primary aluminum. Between 2009 and 2015, China's annual production of primary aluminum increased from 13.68 Mt to 31.52 Mt, with an annual growth rate of approximately 14.9%[39]. In 2022, China's primary aluminum production reached 40.43 Mt, accounting for 59% of the world's total primary aluminum production.

From the regional layout of China's aluminum electrolysis industry, the main producing areas of primary aluminum in China are Shandong, Xinjiang, and Inner Mongolia, which account for 57% of the country's total production. Shandong, Xinjiang, and Inner Mongolia are also the main distribution areas of large primary aluminum plants in

China[40].

The geographical distribution of aluminum electrolysis plants in China has undergone significant changes compared to a decade ago, with Henan, Gansu, and Qinghai being the main raw aluminum producing areas in China more than a decade ago. Private enterprises in provinces such as Shandong and Xinjiang have been vigorously building their own thermal power plants since the early 21st century, and the electricity generated is directly supplied to large aluminum electrolysis plants nearby, forming a coal electricity aluminum industry chain and large-scale aluminum electrolysis production capacity. However, aluminum electrolysis plants that originally relied mainly on grid power in Henan, Gansu, Qinghai, and other places were gradually reduced or even eliminated due to their much higher electricity prices and lack competitiveness compared to their own electricity prices.

It is precisely because of the large-scale construction and expansion of aluminum electrolysis plants that China's primary aluminum production capacity has rapidly increased. In 2022, China's total aluminum electrolysis production capacity reached 45 Mt, and the utilization rate of aluminum electrolysis production capacity was only about 75%.

In recent years, the contradiction of overcapacity in the aluminum industry has become increasingly prominent due to the slowdown in the development speed of the three major areas of domestic aluminum consumption, namely construction, transportation, and electricity. In order to strictly manage and control the new production capacity of primary aluminum, curb the blind expansion of production capacity in the primary aluminum industry, and avoid a serious overcapacity crisis, China has implemented a structural reform on the supply side of primary aluminum since 2017. It is strictly prohibited to add new production capacity for primary aluminum domestically, and actions have been taken to replace and eliminate outdated production capacity.

In 2017, the four national ministries jointly launched a special action to clean up and rectify illegal and irregular projects in the primary aluminum industry, marking the beginning of the replacement of production capacity in the primary aluminum industry. The direction of production capacity transfer is mainly from areas with higher electricity prices to areas with lower electricity prices or abundant green energy. In 2018, the state issued a notice on the implementation of capacity replacement by primary aluminum enterprises through mergers and acquisitions (Ministry of Industry and Information Technology[2018] No. 12), which continuously accelerated the capacity replacement of the domestic aluminum electrolysis industry and further optimized the industrial structure and regional layout. The total amount of cross provincial capacity replacement and transfer in the aluminum electrolysis industry in 2018 was approximately 2.72 Mt, with the main capacity being transferred from Shandong, Henan, Qinghai and other provinces (regions) to Inner Mongolia, Guangxi, Yunnan and other three provinces (regions).

After three years of capacity replacement and elimination, the blind expansion of primary aluminum production capacity has been effectively curbed. Backward and uncompetitive production capacity is rapidly exiting, and the proportion of production capacity using clean energy has increased. The industrial structure and regional layout have been optimized. According to statistics, as of the end of 2019, the cumulative replacement primary aluminum production capacity index is about 10 Mt.

The withdrawal of production capacity is mainly concentrated in Henan Province and Shandong Province, with indicator outflows of 2.87 Mt/a and 2.56 Mt/a, respectively. The production capacity indicators are mainly transferred to provinces such as Inner Mongolia and Yunnan, with a total inflow of 3.84 Mt/a and 3.51 Mt/a, respectively.

The main reason for the large-scale influx of aluminum electrolysis production capacity into Inner Mongolia is that Inner Mongolia has abundant coal resources, low electricity prices after self generation, and the close distance between Inner Mongolia and domestic alumina plants can reduce logistics costs. Therefore, the construction of aluminum electrolysis plants in Inner Mongolia has high competitiveness.

The construction of the hydroelectric aluminum industry chain is an important part of the new round of industrial transfer. With regions such as Yunnan and Sichuan vigorously encouraging the use of clean energy such as hydropower in the aluminum electrolysis industry, there is a trend of transferring aluminum electrolysis production capacity in North China to areas that use green energy such as hydropower. Yunnan and Sichuan, which are rich in hydropower, may become gathering places to undertake a new round of aluminum electrolysis capacity transfer.

1.2.2 Current status of primary aluminum production technology in China

The independent innovation of China's aluminum electrolysis industry has achieved remarkable results in the past thirty years, and overall has reached the world's advanced technological level. This is mainly manifested as: significant breakthroughs have been made in the design and production technology of large electrolytic cells, with basic mastery of the magnetic field, thermal field balance, numerical simulation and design technology of cell structure for large pre baked electrolytic cells[41]. The world's largest 600 kA aluminum electrolytic cell technology has achieved industrialization; The significant energy-saving technology of the new cathode structure aluminum electrolysis cell has been widely promoted and applied; Advanced computer control technology based on alumina concentration and thermal balance control has also been widely applied; We have developed and applied key technologies to improve the lifespan of aluminum electrolysis cells, as well as production technologies for high-quality pre baked carbon anodes and high-quality cathode carbon blocks, and have achieved significant energy-saving and emission reduction effects[42].

The current mainstream technology in China's aluminum electrolysis industry is the use of large-scale pre baked aluminum electrolysis cell technology and advanced computer control systems. Based on the results of physical field simulation research, China has successively independently developed and successfully developed 280 kA, 300 kA, 400 kA, 500 kA and 600 kA large/ultra large prebaked anode aluminum electrolysis cell technologies[43].

At present, China has completely eliminated the outdated and severely polluted 60 kA self baking cell technology and small pre baking cell technology below 160 kA. The production capacity of aluminum electrolysis cells with a capacity of over 300 kA in China has reached 88%, and the production capacity of large aluminum electrolysis cells with a capacity of over 400 kA has reached 70%[44].

China's aluminum electrolysis industry has made significant progress in the development and application of energy-saving, consumption reduction, and clean production technologies. Especially in the past 20 years, a series of energy-saving technologies with independent intellectual property rights in China have been developed and applied. The development and industrial application of new cathode structure aluminum electrolysis cell technology, low-temperature and low-voltage aluminum electrolysis technology, aluminum electrolysis cell fuzzy control technology, petroleum coke coupling and mixing, and high-quality carbon anode production technology have further improved the key technical indicators such as operational stability and DC power consumption of aluminum electrolysis cells in China under low cell voltage.

The comparison of typical technical indicators of various major aluminum electrolysis cells in China with the world's leading technical level of the Hydro 420 kA series aluminum electrolysis cells is shown in Table 1-1.

Table 1-1 Typical technical and economic indicators of 240 kA, 400 kA, and 500 kA aluminum electrolytic cells in China

Parameters	240 kA	400 kA	500 kA	World leading indicators
Current intensity(kA)	240	400	500	420
Current density(A/cm^2)	0.73	0.82	0.8	0.88
Cell voltage(V)	3.99-4.01	3.95	3.9-4.0	4.09-4.12
Current efficiency(%)	91-93	93.5	91-92	95
DC power consumption(Al)(kW·h/t)	12775-13320	12600	12700-13000	12830-12930
Effect coefficient(times/(cell·day))	0.05-0.2	0.05	0.05	0.02-0.07

Note: The international indicator data adopts the world leading level indicators of the Hydro 420 kA series electrolytic cells.

From Table 1-1, it can be seen that the cell voltage and DC power consumption of aluminum electrolysis cells in China are better than the indicators of Hydro 420 kA aluminum electrolysis cells, but the current efficiency is lower than the leading level in foreign countries. In 2022, the average comprehensive AC power consumption of aluminum electrolysis in China has decreased to 13555 kW·h/t, which is about 655 kW·h/t lower than the world average, and the average power consumption has reached the world's leading level.

The most important key technologies for energy conservation, consumption reduction, and clean production in China's aluminum electrolysis industry are described as follows[45]:

(1) Simulation technology for physical field of aluminum electrolysis cell. Over the past 20 years, the aluminum electrolysis technology community in China has conducted in-depth and systematic research on the simulation and simulation of the physical field of aluminum electrolysis cells. They have basically mastered the coupling analysis and simulation of multiple physical fields such as electricity magnetism, electricity heat, heat force, magnetic flow, etc., providing a solid theoretical basis and technical guidance for the development of large and ultra large aluminum electrolysis cell technology in China and the optimization of aluminum electrolysis cell performance indicators, ensuring the stability of the operation of large aluminum electrolysis cells, It has reached the world's advanced level.

(2) Energy saving technology of new cathode structure aluminum electrolysis cell. The new structure aluminum electrolysis cell technology adopts a new cathode structure (including carbon cathode and cathode steel rod structure) and an aluminum electrolysis cell insulation lining design. By reducing the cell voltage and maintaining the thermal

balance and stable operation of the aluminum electrolysis cell, it achieves low electrode distance, low voltage (3.8–3.95 V), low energy consumption (DC power consumption below 12500 kW·h/t), low effect coefficient (less than 0.1), and high energy utilization rate of the aluminum electrolysis production process, This provides important technical support for China's aluminum electrolysis industry to achieve world leading energy consumption.

(3) Intelligent control and manufacturing technology for large prebaked aluminum electrolytic cells. The control system of large-scale pre baked aluminum electrolysis cells in China has evolved from simple cell resistance control to comprehensive control of alumina concentration and thermal balance, from single parameter control to multi parameter coupling control, reaching the world's advanced technological level. The advanced control system has greatly improved the stable operation level of aluminum electrolysis cells in China, improved current efficiency, and reduced various consumption and operating costs. By rapid prediction and fuzzy control of alumina concentration, the anode effect coefficient has been significantly reduced and the service life of aluminum electrolysis cells has been extended.

In the near future, China's aluminum electrolysis plants will generally promote the development of intelligent manufacturing technology and the construction of smart factories, achieving automation, informatization, and intelligence in the detection, control, and management of aluminum electrolysis production processes.

(4) Key technologies for improving the life of aluminum reduction cells. The technology of extending the lifespan of aluminum electrolysis cells has been widely promoted and applied in China. The key technologies for extending the lifespan of aluminum electrolysis cells mainly include: Maintaining thermal balance and stable operation of the cells through advanced operation and control technologies; Adopting high-quality furnace building materials such as new side lining materials and high-quality cathode carbon blocks, as well as improving furnace building quality and optimizing roasting start-up technology, to reduce the early or mid-term damage rate of aluminum electrolysis cells; By using alumina concentration prediction and control technology, the effect coefficient is significantly reduced, and the temperature fluctuation amplitude of aluminum electrolysis cells is reduced; Strengthen detection, timely diagnosis, and handling of aluminum electrolysis cells with longer operating cycles.

The technology of extending the lifespan of aluminum electrolysis cells has increased the average lifespan of aluminum electrolysis cells in China from 1200 d 10 years ago to about 2600 d now, significantly reducing the discharge of waste cell liners, saving billions

CNY in annual maintenance costs for electrolysis cells, and reducing the discharge of nearly 1 Mt of toxic waste cell liners.

1.3 Solid/Hazardous waste disposal in China's aluminum smelting industry

1.3.1 Current situation of red mud discharge and treatment

Red mud is an alkaline solid waste discharged during the production of alumina. Due to differences in bauxite grade, production methods, and technical level, there are significant differences in the yield of red mud[46]. As of the end of 2022, the cumulative storage of red mud in China has exceeded 1.5 billion tons, covering an area of over 120000 acres. The total amount of red mud discharged by the national alumina industry in 2022 was approximately 110 Mt.

The distribution of red mud storage in China is mainly concentrated in provinces and regions such as Shandong, Henan, Guangxi, Shanxi, Guizhou, Yunnan, where alumina plants are built.

The stockpiling of red mud not only wastes land resources, but chemical components such as caustic soda in red mud can also seep into soil and groundwater, and even cause dam collapses and landslides, seriously damaging ecology and polluting the environment.

At present, the vast majority of red mud in China's alumina industry is treated by constructing dams for dry storage. Under natural conditions, different types of cracks may appear in the red mud dam body, even leading to landslides or dam collapse accidents, which will cause serious harm to residents, fields, infrastructure, and ecological environment around and downstream of the red mud storage yard. Fig. 1-3 show the results of red mud dam failures that occurred in Hungary in 2010 and Henan in China in 2016.

At present, alumina plants in China have established an online monitoring and alarm system for the surrounding environment of red mud storage yards, in order to timely grasp the operating status and leakage status of red mud warehouses. This is an important means of preventing disasters and determining treatment plans. The monitoring of the red mud storage yard mainly includes: reservoir water level monitoring, infiltration status line monitoring, dam deformation and displacement monitoring, etc. Establishing an online monitoring system based on computer and information technology can quickly and accurately detect the leakage status of red mud storage yards, reducing the risk of red mud dam failure.

Fig. 1-3 Red mud dam break accident

(a) Hungary red mud reservoir dam break in October 2010; (b) Henan red mud dam landslide in August 2016

The comprehensive utilization rate of red mud in the domestic and foreign alumina industry is very low. The comprehensive utilization rate of red mud in China is only about 5%, far lower than the comprehensive utilization level of other bulk industrial solid waste in China. China Aluminum Shandong Branch currently has the highest utilization rate of red mud among Chinese alumina enterprises, reaching around 25%. The strong alkalinity, complex composition, and trace radioactive elements of red mud are the main reasons for its difficulty in treatment and large-scale utilization. The high cost of comprehensive utilization of red mud is also another reason for limiting its utilization.

The metal iron content in high iron red mud ranges from 30% to 50%, and the iron minerals in it are mainly weakly magnetic hematite and goethite. At present, China has developed the use of high gradient strong magnetic separation technology to separate iron concentrate from high iron red mud. China Aluminum Guangxi Branch's high-speed iron red mud magnetic separation can obtain iron concentrates with iron metal content higher than 55%, and the iron recovery rate is greater than 22%. This technology is more suitable for the treatment of high iron red mud using the high-temperature Bayer process in China, but the iron recovery rate is still relatively low. At present, methods such as reduction roasting magnetic separation and smelting pig iron are still being studied. The main problem that needs to be solved is to improve the recovery rate of iron in red mud and the grade of iron concentrate, while reducing the energy consumption and cost of the treatment process.

The large-scale production of building materials from red mud is the main direction for the comprehensive utilization of red mud. The use of red mud to prepare unburned bricks, permeable bricks, building bricks, cement and microcrystalline glass, and other building materials, or as filling materials for embankments or plastics, is a practical example of using red mud as a building material. At present, the production of unburned bricks, permeable bricks, building bricks, cement, and road construction using red mud has achieved small-scale industrial applications. The technology for producing microcrystalline glass and plastic fillers is still being studied. The main problem with using red mud as a building material is that its strong alkalinity and radioactivity make the produced building materials also contain certain alkalinity and radioactivity, making it difficult to meet the standards of building materials.

Another comprehensive utilization direction of red mud is to utilize its strong alkalinity to neutralize acidic wastewater, sludge, and exhaust gas, thereby achieving waste treatment. The main application areas are used to treat acidic wastewater from mineral processing and metallurgy, urban river sludge, and flue gas containing sulfur dioxide, but there are currently no detailed reports or evaluation materials on actual utilization. Red mud treatment has been applied on a small-scale scale to improve acidic soil. The direction of red mud utilization should become an important field for future red mud utilization.

At present, the treatment method of red mud, the waste residue from alumina production in China, is dry stacking and safety monitoring technology is used to achieve the safe disposal of red mud. The main problems currently existing are the safety of red mud dam storage and the infiltration pollution of alkaline solution. Due to the huge production of alumina in China and the continuous decline in the grade of bauxite, a large amount of red mud discharged from alumina production will still be produced, and the impact of red mud on the environment will be even more severe. In addition, alumina plants in China are often densely constructed and not far from residential areas and red mud yards, making the impact on the environment even more worrying.

The urgent task is to build a safe and reliable red mud yard, resolutely avoid leakage or dam collapse, and eliminate secondary pollution. It is necessary to comprehensively promote the application of red mud dry stacking technology, prevent wet stacking, and develop new red mud pressure filtration technology to minimize the amount of alkali entering the storage yard. Improve the anti-seepage level and online safety detection level of the red mud yard to completely prevent the leakage and collapse of the red mud alkali liquor.

It is necessary to intensify research on anti-seepage, alkali reduction, and reclamation technologies after the closure of the red mud reservoir, improve the green coverage rate of the red mud reservoir, and turn the original red mud reservoir into grassland, pasture, and even gradually into arable land.

In the long run, efforts should be made to develop technology for the resource utilization of red mud, and research and development should be carried out in collaboration with large-scale industries such as building materials, steel, chemical engineering, and environmental protection to treat red mud. Developing key technologies for harmless, reducing, and resource utilization of red mud waste residue, solving the problem of resource utilization for large-scale consumption of red mud, minimizing the pollution impact of the alumina industry on the environment, and promoting the development of circular economy in China's alumina industry and the construction of resource saving enterprises.

1.3.2 Flue gas and wastewater discharge and treatment

The high-temperature flue gas emitted during the production of alumina mainly comes from high-temperature kilns such as boilers, gas furnaces, aluminum hydroxide calciners, and lime kilns. The flue gas mainly contains gases such as CO_2, SO_2, NO_x, water vapor, and dust particles.

At present, equipment such as electrostatic precipitators and bag filters are commonly used to remove dust from flue gas, in order to reduce the concentration of dust in the exhaust gas. Simultaneously developing lime gypsum or ammonia desulfurization technology, flue gas catalytic denitrification technology, flue gas heat exchange recovery waste heat and water vapor technology, etc., can basically achieve the standard emission of flue gas pollutants in alumina production, but the treatment cost is still high.

The wastewater in the production process of alumina mainly comes from sewage tanks, equipment cooling, and other aspects. By strengthening management, implementing separation of clean and dirty water, combining segmented treatment with centralized treatment, and classifying and recycling wastewater, zero discharge of wastewater in the alumina industry can be basically achieved.

An important environmental pollution of aluminum electrolysis enterprises is the emission of harmful gases during the production process of primary aluminum, including various harmful gases such as hydrogen fluoride, perfluorocarbons, sulfur dioxide, and fluorine containing dust particles.

The fluoride emissions from the aluminum electrolysis production process include hydrogen fluoride(HF) generated by the reaction between water and fluoride salts, and smoke (fluoride dust particles) generated by the volatilization of fluoride salts. The emission pathways include organized emissions and unorganized emissions. Gaseous

1.3 Solid/Hazardous waste disposal in china's aluminum smelting industry

hydrogen fluoride is produced during the electrolysis process and is transported through the flue gas to the flue gas purification treatment centers (GTCs). The vast majority of hydrogen fluoride is adsorbed by alumina and returned to the aluminum electrolysis cell. However, in the electrolysis workshop, anode assembly, residual electrode treatment, cell lining overhaul, electrolyte crushing and other processes, if attention is not paid to the recovery and centralized treatment of fluoride salts, unorganized emissions of fluoride will occur, as shown in Table 1-2.

Table 1-2 Typical fluoride/sulfur dioxide purification and emissions of electrolytic series

No.	Process equipment	HF	Fluorine containing particles	SO_2
1	Maximum discharge of electrolytic cell (kg)	18	10	20
2	Dry purifier recovery (kg)	17.73	9.85	19.7
3	Recycled (secondary) alumina content (kg)	17.68	9.8	9.0
4	Sunroof emissions (kg)	0.27	0.15	0.30
5	Chimney emissions (kg)	0.05	0.05	10.70
6	Ratio of sunroof emissions (%)	84	75	97
7	Proportion of severe emissions (%)	16	25	3

As shown in Table 1-2, environmental pollutants such as fluoride and dust generated during the electrolysis process will become a key source of flue gas pollution for primary aluminum plants if they are not treated by a dry purification system but are discharged unorganized through workshop skylights and other emission channels. According to statistics, this unorganized emission accounts for almost three-quarters of the total emissions of fluoride and particulate matter.

At present, the direction to solve fluoride emission pollution is to improve the purification efficiency of organized flue gas emissions, so that the exhaust gas discharged from the chimney meets the emission standards. Moreover, by improving the sealing coefficient of aluminum electrolysis cells and strengthening the recovery and treatment of various process exhaust gases, the amount of unorganized emissions is minimized as much as possible. It is particularly important to establish strict fluoride emission standards and local total emissions, and to standardize the overall scale and operation management of aluminum electrolysis construction.

At present, China's fluoride emissions per ton of aluminum comply with the 2010 Ministry of Industry and Information Technology regulation of ≤0.6 kg. In addition, the State Environmental Protection Administration has also proposed a requirement that the average concentration limit of fluoride in the air pollutants at existing and newly built boundaries is 0.02 mg/m^3.

The emission of fluoride is related to the fluoride consumption per ton of aluminum in aluminum electrolysis plants. Foreign advanced enterprises control the fluoride emission per ton of aluminum at $0.3-0.35$ kg fluoride/ton of aluminum, even lower than 0.15 kg fluoride/ton of aluminum.

Under different environmental capacities, the construction scale of primary aluminum plants also varies. In low fluorine areas of China, the construction scale of a single primary aluminum plant should not exceed 800 kt/a. It is more reasonable to build a scale of less than 600 kt/a in mountainous areas in the southwest. In areas with low population density such as Xinjiang and Inner Mongolia, 1 Mt of primary aluminum plants can be built. If the comprehensive fluoride discharge reaches the world's advanced level (less than 0.25 kg/t), it is still feasible to achieve a scale of 1.2 Mt/a or more.

Therefore, the setting of the construction scale of aluminum electrolysis plants not only requires attention to electricity prices, supply conditions for alumina and carbon anodes, and competitiveness analysis, but also requires environmental assessment to strictly control the emission of aluminum electrolysis flue gas and fluoride pollution.

1.3.3 Discharge and treatment of anode carbon residue in aluminum electrolysis process

The anode carbon residue generated during the aluminum electrolysis process is an important source of aluminum electrolysis pollution and belongs to solid hazardous waste. In 2022, the amount of anode carbon slag discharged in China reached a scale of about 300 kt, equivalent to approximately $6-10$ kg of anode carbon slag per ton of primary aluminum.

The electrolyte content in anode carbon slag is about 50%, while the rest is carbon particles. Improper disposal of carbon residue not only causes significant losses of fluoride salts, but also causes serious environmental pollution. In 2016, anode carbon residue was included in the "National Hazardous Waste List" (waste code: 321-025-48, hazardous characteristic: T), and it is prohibited to dispose of or store in the open air. Relevant environmental policies require primary aluminum production enterprises to conduct harmless treatment in the factory or entrust units with hazardous waste treatment qualifications for treatment.

At present, the main technologies for treating aluminum electrolysis carbon residue in China include flotation, roasting, and other methods. The flotation method involves grinding the carbon residue with water to meet the required concentration and particle size, adding flotation agents for stirring treatment, and then entering the flotation machine and introducing air to form bubbles. At this time, the floatable materials (mainly carbon powder) will stick to the bubbles and float to form foam, which will be scraped out and separated into carbon powder. The non floatable materials (mainly electrolyte) will be discharged from the bottom flow of the flotation cell, so as to achieve the purpose of separating the electrolyte and carbon powder. The advantages of flotation method are low processing cost and low labor intensity for workers. However, the disadvantage is that the generated electrolyte is in a fine powder form and has a high carbon content (about 5%). Returning it directly to the aluminum electrolysis cell will have a significant impact on electrolysis production. The carbon powder produced by flotation also contains fluoride salts, which are difficult to utilize. In addition, flotation wastewater contains fluoride salts, causing environmental pollution.

The roasting method for treating anode carbon residue is to place the anode carbon residue in a high-temperature furnace and introduce natural gas for roasting. The electrolyte in the carbon residue becomes a molten liquid at high temperature and flows out of the chute at the bottom of the furnace for recycling. Part of the carbon residue undergoes high-temperature treatment and is converted into carbon dioxide and fluorinated gas, which are sent to the gas purification system. The remaining carbon residue is removed and stored by a slag removal truck. The block electrolyte recovered by this method has high purity and can be directly used as an electrolytic raw material. However, high-temperature roasting requires a large amount of natural gas consumption and generates secondary environmental issues, with high labor intensity and harsh environment.

Overall, existing technologies for the comprehensive utilization of anode carbon residue resources have their own advantages and disadvantages, but they have not fully solved the problems of harmless treatment and resource utilization of anode carbon residue.

The future anode carbon residue treatment technology should develop towards a more comprehensive utilization of carbon residue resources, more economical treatment costs, more environmentally friendly process, and easy industrial application.

1.3.4 Emission and treatment of hazardous solid waste

The service life of aluminum electrolysis cells in China is currently generally between 2200 d and 2800 d. During the overhaul of aluminum electrolysis cells, all waste lining materials (including waste cathode carbon blocks, waste refractory materials, and waste silicon carbide side blocks) inside the cells are replaced[47].

The primary aluminum plant will produce 24–30 kg of spent potlining for every 1 t of aluminum produced. In recent years, China has produced approximately 1 Mt of aluminum electrolytic cell waste lining annually. In 2022, approximately 1.07 Mt of spent potlining were produced throughout the year.

The spent potlining contains high levels of water-soluble fluoride and trace amounts of cyanide, which can cause significant harm to the ecological environment and human society if discharged directly. In 2016, the spent potlining was included in the National Hazardous Waste List (waste code: 321-023-48, hazardous characteristic: T), which requires strict control and harmless disposal.

At present, most foreign aluminum electrolysis plants crush all the spent potlining and send them to the cement plant for batching and burning into cement clinker.

The treatment technologies for aluminum electrolysis waste cell liners internationally include Rio Tinto's LCL&L wet process technology and Renault's and Alcoa's pyrometallurgical technology. Wet process technology for oxidative decyanation at 180 ℃-adding lime for fluoride fixation; The pyrometallurgical technology also involves adding lime to fix fluoride and high-temperature oxidation to remove cyanide, both of which produce inert solid waste. Domestic aluminum companies have conducted research on the spent potlining fire treatment technology, but it is difficult to industrialize due to high energy consumption and the generation of new solid waste residue. Beijing Mining and Metallurgy Technology Group and State Power Investment Qingtongxia Aluminum Industry Co., Ltd. have cooperated to build a wet treatment line for waste cathode carbon blocks. However, the pollution during the treatment process is severe, making it difficult to completely separate carbon powder and cryolite.

Since 2017, China has put forward clear requirements for environmental policies for aluminum electrolysis plants. Several domestic aluminum electrolysis plants have built multiple production lines using calcium hypochlorite to remove cyanide and adding calcium chloride to generate soluble fluoride into calcium fluoride. However, this technology has high treatment costs, generates a large amount of wastewater, exhaust gas, and salt containing waste residue during operation, and is still hazardous waste, which has not been successfully applied in industry. Therefore, China urgently needs to develop harmless and resourceful treatment technologies for spent potlining.

Considering that the spent potlining contains high value metallic elements such as graphite carbon, fluoride salts, silicon carbide, aluminum iron, etc., implementing classified treatment of the spent potlining and achieving collaborative disposal with the cement industry, steel industry, mineral wool industry, and power industry is the development direction of realizing the resource utilization technology of the spent potlining.

The properties of waste cathode carbon blocks and silicon carbide are different, and different treatment methods should be used to achieve the separation of fluorine from carbon materials such as carbon or silicon carbide, so that high graphitized carbon and silicon carbide materials can be recycled and utilized. A better low-cost disposal method for waste refractory materials is to collaborate with nearby industries such as cement for treatment.

The aluminum ingot casting process produces hazardous solid waste-aluminum dross, which accounts for approximately 1% - 1.5% of the ingot production[48]. The main components of aluminum electrolysis aluminum dross are metal aluminum, alumina, and aluminum nitride, as well as small or trace amounts of calcium, magnesium, potassium, sodium fluoride, etc. After extracting metal aluminum from primary aluminum dross, waste residue secondary aluminum dross is obtained[49].

The aluminum nitride and various salts in aluminum dross can cause serious pollution to the ecological environment and endanger human health. At the same time, aluminum dross contains a large amount of valuable resources, which, if not utilized, will result in resource waste. Therefore, there is an urgent need to develop clean, environmentally friendly, and economically efficient aluminum dross resource utilization technologies.

Before 2016, aluminum electrolysis aluminum dross had not yet been classified as hazardous solid waste by the state. Most aluminum electrolysis plants only extracted a portion of metal aluminum from primary aluminum dross, and the resulting secondary aluminum dross was sold as waste to surrounding small enterprises. These small enterprises use simple methods to produce water purification agents, steel refining agents, etc., but the treatment process is severely polluted and the product value is low. Most aluminum electrolysis enterprises use landfill to treat secondary aluminum dross, which will inevitably cause environmental pollution.

Foreign countries have developed and applied various technologies to recover metal aluminum from primary aluminum dross, including pressing recovery method, rotary kiln melting recovery method, tilting rotary furnace method, ALUREC method, etc. The recovery rate of metal aluminum can generally reach over 80%. However, for secondary aluminum dross, it is generally only used for waste stacking or landfill treatment[50].

In China, metal aluminum is recovered from primary aluminum dross through heat treatment recovery methods (including roasting ash recovery method and pressing recovery method, etc.). The processing technology and equipment are relatively backward, with low recovery rate, high energy consumption, and serious pollution. Therefore, there is an urgent need for improvement and improvement. There has been a lot of laboratory research on the recycling and utilization of secondary aluminum dross in China, mainly focusing on the preparation of alumina, aluminum sulfate water purifier, steelmaking promoter, building materials, and refractory materials. However, most of these technologies are difficult to achieve industrial scale applications due to their economic infeasibility or the existence of technical, equipment, and environmental challenges.

Overall, there is currently a high level of attention both domestically and internationally on the resource utilization technology for the recovery of metallic aluminum from aluminum dross, which has also been industrialized, but the recovery efficiency needs to be improved. However, the treatment and utilization of secondary aluminum dross remains largely in the laboratory research stage.

Secondary aluminum dross is the biggest challenge for the safe disposal and resource utilization of aluminum dross, and it is also the main source of environmental pollution. We should develop green and clean resource utilization technologies for secondary aluminum dross as soon as possible to achieve sustainable development of China's aluminum industry.

1.4 Sustainable development strategies for China's aluminum smelting industry

1.4.1 Sustainable development strategies for China's alumina industry

In order to ensure a stable and reliable supply of bauxite resources for China's alumina industry, reasonably utilize domestic and foreign resources, and eliminate the bottleneck of sustainable development of the alumina industry, we should focus on the stable and orderly mining of existing bauxite, strictly and reasonably control the mining output of each ore spot, and ensure the safety and environmental protection of the mining process. On the basis of mature development of corresponding technologies, gradually mining complex composition of medium and low grade bauxite, with a focus on high-grade diaspore ore and high sulfur diaspore ore under coal seams, to expand the source of bauxite resources. Deeply study the development strategy of stable and reliable access to overseas bauxite. With the help of the "the Belt and Road" initiative and the "community with a shared future for mankind" policy, adopt the internationally unified form of acquisition, holding or equity participation to stabilize access to overseas high-quality bauxite resources. Build large alumina plants using imported mines and energy in appropriate coastal areas of China. Choose to build alumina plants in overseas areas with abundant bauxite resources.

According to the systematic research on the production and consumption of primary aluminum in China, it can be determined that the output of China's aluminum industry has reached the top platform area. The existing production capacity of alumina and primary aluminum can meet domestic demand for a considerable period of time in the future. Therefore, the production capacity of alumina in China should be controlled within 84 Mt, and the annual output of alumina should be limited to 72 – 75 Mt. The construction capacity of new alumina plants in the future should be matched with the elimination of old plant capacity to maintain stable and controllable alumina production capacity and output in China.

Alumina enterprises should optimize their production capacity layout. Alumina plants in northern regions with declining bauxite grade and frequent resource depletion should consider transferring their alumina production capacity to coastal areas as soon as possible to facilitate the production of alumina using imported mines. Alumina plants in Guangxi, Yunnan, and Guizhou provinces should allocate reasonable production capacity based on the reserves of bauxite resources to prevent blind and disorderly expansion. The newly built alumina plants along the coast must ensure sufficient supply of overseas bauxite resources. If conditions permit, it is possible to selectively build competitive alumina plants overseas.

The selection of the location for constructing an alumina plant should also consider whether it is convenient to transport the product alumina to the primary aluminum plant. If possible, an alumina primary aluminum joint venture can be established to save on the transportation cost of alumina. The suitable locations for constructing coastal alumina plants can generally be in Guangxi (available for aluminum electrolysis plants in Yunnan and Guangxi), southern Shandong and northern Jiangsu (available for aluminum electrolysis plants in central and western China), and northern Inner Mongolia or Liaoning (available for aluminum electrolysis plants in Inner Mongolia and Liaoning).

Relying on technological progress, we aim to address the efficient, energy-saving, and low-cost production of alumina from complex diaspore mines in China. We aim to develop and apply technologies for producing alumina from high silicon, high sulfur, high lithium, and high organic bauxite, ensuring that China's alumina industry has a certain amount of domestically produced bauxite resources that can be economically utilized, in order to reduce the cost of bauxite.

Relying on technological progress, we have developed technologies for deep energy conservation and clean production of alumina, especially to further improve cycle efficiency and waste heat utilization technology, so that China's alumina production energy consumption can enter the world's advanced level and improve its core competitiveness.

Relying on technological progress to achieve automation, informatization, and intelligence in alumina production. In a relatively short period of time, establish corresponding demonstration enterprises to enable China's alumina industry to catch up with the world's advanced level in intelligence and achieve sustainable development.

The environmental strategy for the sustainable development of China's alumina industry focuses on the safe disposal and resource utilization of red mud.

The existing dry storage technology of red mud must be comprehensively promoted and further optimized. The first is to improve the dry storage technology, including the technology of pressure filtration and distribution of red mud; The second is to strengthen the anti-seepage, online monitoring, and safety management of the storage yard, to minimize the adverse impact of the red mud storage yard on the environment to the greatest extent; The third is to gradually promote greening and reclamation after the closure of red mud storage yards.

The resource utilization of red mud should be coordinated with large-scale industries such as building materials, steel, chemicals, and environmental protection to produce red mud based building materials, auxiliary materials for the steel and chemical industries, and environmentally friendly materials for treating acidic waste. This will solve the problems in the technological development and production management of large-scale consumption and resource utilization of red mud, achieve the reduction, harmless, and resource utilization of red mud waste, and build a green environment sustainable development of China's alumina industry.

1.4.2 Sustainable development strategies for China's aluminium reduction industry

China's annual production of primary aluminum accounts for about 56% of the world's total production, and it is necessary to strictly control the total production capacity and output of aluminum electrolysis in China. According to the estimation of domestic demand for primary aluminum in recent years, it is advisable to control the annual production of primary aluminum at around 36 Mt.

Under the condition of controlling total production, implement orderly transfer of production capacity and eliminate outdated ones. Eliminate electrolytic production capacity with high electricity consumption, high electricity prices, and low efficiency, transfer production capacity to green energy areas, low electricity price areas, and stable power supply areas, scientifically coordinate power supply prices, and gradually solve the fairness and rationality of electricity prices for aluminum electrolysis.

The main characteristics of aluminum electrolysis technology in China are low voltage, low power consumption, and low efficiency. The main key technical problem to be solved is to improve current efficiency and production efficiency under low voltage conditions.

Therefore, the technological progress in improving the core competitiveness of aluminum electrolysis in China in the future will mainly be reflected in the following aspects:

(1) By further optimizing the physical field design technology of large capacity aluminum electrolysis cells, which is characterized by the stability of magnetic fluid in the electrolysis cell, and adopting measures such as high-quality carbon anodes, sandy alumina, and graphitized cathode carbon blocks, the current efficiency of aluminum electrolysis can be improved under low cell voltage and higher current density operating conditions, achieving further energy conservation and consumption reduction.

(2) Develop intelligent process control technology to reduce the anode effect coefficient, improve the stability of the aluminum electrolysis process, and improve production efficiency.

(3) Vigorously promote the application of high-quality carbon anode production technologies such as petroleum coke coupling and mixing, prioritize the application of domestically produced high-quality petroleum coke and carbon anode in domestic aluminum electrolysis plants, reduce carbon slag production rate, maintain the stability of the aluminum electrolysis process, and achieve energy conservation, consumption reduction, and emission reduction.

(4) Develop an intelligent control system for aluminum electrolysis, achieving standardization, visualization, automation, and intelligence of the aluminum electrolysis production process operation. Strictly control the superheat of the aluminum electrolyte, maintain the optimization and stability of the furnace shape and structure of the reduction cell, reduce bottom sedimentation and shell formation, reduce horizontal current, and improve current efficiency and production efficiency.

It is necessary to strictly establish emission standards and control regulations for aluminum electrolysis flue gas, especially to solve the measurement and estimation standards for unorganized emissions of flue gas, technically solve the problems of unorganized emissions and efficient purification of flue gas in aluminum electrolysis, achieve efficient purification and standard emissions of fluorine and sulfur in flue gas of aluminum electrolysis plants, and gradually establish local total emission control standards.

Developed safe disposal and high-value resource utilization technologies for aluminum electrolysis waste such as anode carbon slag, waste cathode carbon blocks, waste silicon carbide side lining, waste refractory materials, and aluminum dross.

Implement classified management and disposal of various solid waste residues from aluminum electrolysis, and develop optimal treatment technologies for different types of waste as soon as possible. The goal is to make reasonable and high-value use of valuable components in waste, making China's aluminum electrolysis industry a green and sustainable modern industry without hazardous waste residue discharge.

References

[1] PAN Z S, ZHANG Z Z, ZHANG Z N, et al. Analysis of the import source country of the bauxite in China[J]. China Mining Magazine, 2019, 28(2):13-17.

[2] GAO X T, GUO S, LI H L. Summary and study of the imported bauxite in China[J]. Light Metals, 2016, (7):4-11.

[3] ZHAO B Y, MEN C S. Discussion on bauxite supply prospect for domestic alumina refineries[J]. Light Metals, 2016, (8):8-12.

[4] MA D C, XU Y M, GUO C Q. Analysis on the current situation of using imported bauxite in domestic alumina plants[J]. China Nonferrous Met., 2019(1)1:50-52.

[5] Ministry of Natural Resources, PRC. China Mineral Resources[M]. Beijing: Geological Publishing House, 2019.

[6] DILLINGER B, BATCHELOR A, KATRIB J, et al. Microwave digestion of gibbsite and bauxite in sodium hydroxide[J]. Hydrometallurgy, 2020, 192:105257.

[7] AKINCI A, ARTIR R. Characterization of trace elements and radionuclides and their risk assessment in red mud[J]. Mater Charact., 2008, 59:417-421.

[8] HUANG Y F, HAN G H, LIU J T, et al. A facile disposal of Bayer red mud based on selective flocculation desliming with organic humics[J]. J. Hazard. Mater., 2016, 301:46-55.

[9] PAN X L, YU H Y, TU G F. Reduction of alkalinity in bauxite residue during Bayer digestion in high-ferrite diasporic bauxite[J]. Hydrometallurgy, 2015, 151:98-106.

[10] QIAN Y, LI S S, CHEN X L. Synthesis and characterization of LDHs using Bayer red mud and its flame-retardant properties in EVA/LDHs composites[J]. J. Mater. Cycles. Waste., 2015, 17:646-

654.

[11] WANG Y, LI X, ZHOU Q, et al. Observation of sodium titanate and sodium aluminate silicate hydrate layers on diaspore particles in high-temperature Bayer digestion[J]. Hydrometallurgy, 2020, 192: 105255.

[12] SILVEIRA L M V, MACIEL M H, MESQUITA J A F S, et al. Hydration and chemical shrinkage of Portland cement with bauxite residue from Bayer process[J]. Journal of Thermal Analysis and Calorimetry, 2023, 148(14): 6715-6729.

[13] ZHOU G T, WANG Y L, ZHANG Y G, et al. A clean two-stage Bayer process for achieving near-zero waste discharge from high-iron gibbsitic bauxite[J]. Journal of Cleaner Production, 2023, 405: 136991.

[14] BIAN J W, LI S, ZHANG Q L. Experimental investigation on red mud from the Bayer process for cemented paste backfill[J]. International Journal of Environmental Research and Public Health, 2022, 19(19): 11926.

[15] ZHANG N, NGUYEN V A, ZHOU C. Impact of interfacial Al- and Si-active sites on the electrokinetic properties, surfactant adsorption and floatability of diaspore and kaolinite minerals [J]. Minerals Engineering, 2018, 122: 258-266.

[16] ZHANG N, ZHOU C, LIU C, et al. Effects of particle size on flotation parameters in the separation of diaspore and kaolinite[J]. Powder Technology, 2017, 317: 253-263.

[17] CHENG L W, WANG Y L, ZHOU Q S, et al. Scale formation during the Bayer process and a potential prevention strategy[J]. Journal of Sustainable Metallurgy, 2021, 7(3): 1293-1303.

[18] JHA V K, AGNIHOTRI A, SHARMA R J, et al. Mathematical model for estimating variation in specific energy consumption with respect to capacity utilization for aluminum smelting plant[J]. Energy Efficiency, 2018, 11(3): 773-776.

[19] DING J, MA S, SHEN S, et al. Research and industrialization progress of recovering alumina from fly ash: A concise review[J]. Waste Manag., 2017, 60: 375-387.

[20] VALEEV D, KUNILOVA I, ALPATOV A, et al. Complex utilisation of ekibastuz brown coal fly ash: Iron & carbon separation and aluminum extraction[J]. J. Clean. Prod., 2019, 218: 192-201.

[21] LI Y, ZHANG Y, YANG C, et al. Precipitating sandy aluminium hydroxide from sodium aluminate solution by the neutralization of sodium bicarbonate[J]. Hydrometallurgy, 2009, 98: 52-57.

[22] LI X B, YAN L, ZHAO D F, et al. Relationship between $Al(OH)_3$ solubility and particle size in synthetic Bayer liquors [J]. Trans. Nonferrous Met. Soc. China (Engl. Ed.), 2013, 23: 1472-1479.

[23] WANG Y, ZHANG T, LYU G, et al. Recovery of alkali and alumina from bauxite residue (red mud) and complete reuse of the treated residue[J]. J. Clean. Prod., 2018, 188: 456-465.

[24] GU F Q, LI G H, PENG Z W, et al. Upgrading diasporic bauxite ores for iron and alumina enrichment based on reductive roasting[J]. JOM, 2018, 70(9): 1893-1901.

[25] SENYUTA A, PANOV A, SUSS A, et al. Innovative technology for alumina production from low-grade raw materials[J]. Light Metals, 2013: 203-208.

[26] CHEN J, LI X, CAI W, et al. High-efficiency extraction of aluminum from low-grade kaolin via a

[27] LI G, YE Q, DENG B, et al. Extraction of scandium from scandium-rich material derived from bauxite ore residues[J]. Hydrometallurgy, 2018, 176:62-68.

[26] novel low-temperature activation method for the preparation of poly-aluminum-ferric-sulfate coagulant[J]. J. Clean. Prod. ,2020,257:120399.

[28] PANIAS D, ASIMIDIS P, PASPALIARIS I. Solubility of boehmite in concentrated sodium hydroxide solutions: Model development and assessment[J]. Hydrometallurgy, 2001, 59:15-29.

[29] LIU G, LI Z, QI T, et al. Two-Stage process for precipitating coarse boehmite from sodium aluminate solution[J]. JOM, 2017, 69:1888-1893.

[30] LI H, ADDAI-MENSAH J, THOMAS J C, et al. The influence of Al(Ⅲ) supersaturation and NaOH concentration on the rate of crystallization of Al(OH)$_3$ precursor particles from sodium aluminate solutions[J]. J. Colloid Interface Sci. ,2005,286:511-519.

[31] ALEX T C, KUMAR R, ROY S K, et al. Towards ambient pressure leaching of boehmite through mechanical activation[J]. Hydrometallurgy, 2014, 144:99-106.

[32] LI X B, FENG G T, ZHOU Q S, et al. Phenomena in late period of seeded precipitation of sodium aluminate solution[J]. Trans. Nonferrous Met. Soc. China(Engl. Ed.) ,2006,16:947-950.

[33] WIND S, RAAHAUGE B E. Experience with commissioning new generation gas suspension calciner[J]. Miner. Met. Mater. Ser. ,2016:155-162.

[34] LIU F Q, LIU Y X, MANNWEILER U, et al. Effect of coke properties and its blending recipe on performances of carbon anode for aluminium electrolysis[J]. J. Central South Univ. Technol. , 2006,13(6):647-652.

[35] LIU F Q, YANG H J, YANG X P. Industrial production and application of high quality cathodes in China[J]. Carbon Tech. ,2007,26(4):36-40.

[36] LI J. Improvement of anode of prebaked cell in aluminum electrolysis[J]. Advances in Material Science, 2021, 5(1):1-3.

[37] LIU F Q, XIE M Z, LI R B, et al. Impurity migrations in aluminum reduction process and quality improvement by anti-oxidized prebaked anode[J]. Journal of Sustainable Metallurgy, 2021, 7 (2):427-436.

[38] AZARI K, ALAMDARI H, ARYANPOUR G, et al. Mixing variables for prebaked anodes used in aluminum production[J]. Powder Technology, 2013, 235:341-348.

[39] LIU F Q, LV H, ZUO Z P, et al. A denitrification-phase transition and protection rings(DPP) process for recycling primary aluminum dross[J]. ACS Sustainable Chemistry & Engineering, 2021,9(41):13751-13760.

[40] CULLEN J M, ALLWOOD J M. Mapping the global flow of aluminum: From liquid aluminum to end-use goods[J]. Environ. Sci. Technol. ,20143,47:3057-3064.

[41] VICKY V, ANTONIS P, MARIA T. A mathematical and software tool to estimate the cell voltage distribution and energy consumption in aluminium electrolysis cells[J]. Materials Proceedings, 2022,5(1):116-121.

[42] CUI H, XIE G, CHEN S R, et al. Application of intelligent control in aluminium cell[J]. J. Kunming Univ. Sci. Technol. ,2001,26(6):101-106.

[43] LI J,LV X J,LAI Y Q,et al. Research progress in TiB$_2$ wettable cathode for aluminum reduction [J]. JOM,2008,60(8):32-37.

[44] MUNGER D, VINCENT A. Electric boundary conditions at the anodes in aluminum reduction cells[J]. Metallurgical and Materials Transactions B. ,2006,37(6):1025-1035.

[45] ARKHIPOV V G. The mathematical modeling of aluminum reduction cells[J]. JOM,2006,58(2):54-56.

[46] YIN Y Y,KONG L S,YANG C H,et al. Optimal operation of alumina proportioning and mixing process based on stochastic optimization approach [J]. Control Engineering Practice, 2021, 113:104855.

[47] PARASKEVAS K,KELLENS A,VANDE VOORDE W,et al. Environmental impact analysis of primary aluminium production at country Level[J]. Procedia CIRP,2016,40:209-213.

[48] GIL A. Management of the salt cake from secondary aluminium fusion processes[J]. Ind. Eng. Chem. Res. ,2005,44:852-8857.

[49] DAVID E,KOPAC J. Hydrolysis of aluminum dross material to achieve zero hazardous waste[J]. J. Hazard. Mater. ,2012,209-210:501-509.

[50] MAHINROOSTA M,ALLAHVERDI A. Hazardous aluminum dross characterization and recycling strategies:A critical review[J]. J. Environ. Manage. ,2018,223:452-468.

Chapter 2   Spent Potlining Disposal Technology

2.1 Overview

With the rapid development of the aluminum reduction industry, the amount of carbon dust, aluminum dross and spent potlining generated in the aluminum reduction process has increased year by year. It is estimated that 8–12 kg of carbon dust, 10–12 kg of aluminum dross, and 20–30 kg of spent potlining will be produced for each ton of aluminum produced[1]. In 2022, a total of 365 kt of carbon dust, 395 kt of aluminum dross and 900 kt of spent potlining were produced in China. The spent potlining consists of three parts: Spent cathode carbon block, waste silicon carbide side block and waste refractory material.

All of the above materials contain high value aluminum, alumina, carbon, silicon carbide, fluoride salts, and cryolite. However, they are all complex mixtures with chemical composition. In particular, the lining of the reduction cell is eroded and infiltrated by the high temperature electrolyte during the long-term production process. Therefore, the cathode carbon block, the silicon carbide side block and the refractory material generated after the overhaul of the reduction cell contain a large amount of soluble fluoride salt and a small amount of cyanide[2]. Soluble fluorine and cyanide in these substances can be dissolved into the soil and groundwater while releasing harmful gases when they are stacked outdoors and in contact with humid air or rain. It will cause great harm to the local environment and society. At present, these aluminum reduction industrial wastes have been listed in the *National Hazardous Waste List* in China.

The huge amount of industrial hazardous waste is not only a great burden on the environment, but also a serious waste of valuable resources. Scholars from various countries are exploring ways to clean and efficiently treat these hazardous wastes.

However, the chemical composition and phase composition of these hazardous solid wastes are complicated. There are also many newly formed compounds that are tightly fused and embedded with each other, making separation extremely difficult. Although a large number of experimental research works have been carried out in the world, various treatment technologies have been proposed, but most of the aluminum reduction plants have not treated these hazardous solid wastes due to costs, technical maturity and the like.

In recent years, a variety of solid pollutants in the aluminum reduction industry have been listed in the *National Hazardous Waste List* in China. Under the pressure of this policy, the aluminum factories have selected some treatment technologies and built several processing lines. The method of flotation is used to treat the anode carbon residue, and the calcium element is used to fix the fluorine. However, these treatment technologies all have drawbacks such as large investment, poor product quality, high operating cost, and secondary pollution. To this end, a series of key technologies and equipment for the clean treatment and resource utilization of hazardous wastes in the aluminum electrolysis industry are urgently needed. The basic theoretical research on hazardous waste in the aluminum reduction industry is also necessary.

2.1.1 Cement industry

The use of waste cathode carbon blocks and waste refractory materials in the cement manufacturing industry has been reported[3-5]. The composition of cement is $CaO-SiO_2-Al_2O_3-Fe_2O_3$ system, and the carbon in the waste cathode carbon block serves as a supplementary fuel in cement manufacturing. The alkali metal fluoride in it can serve as a catalyst in the sintering reaction of the furnace material. Therefore, the sintering temperature of the clinker can be reduced and the fuel consumption can be reduced. The combustion method of adding waste cathode carbon blocks as fuel in dry process cement kilns can also be applied in the cement industry. Usually, cement kilns use coal powder as fuel, and the finely ground waste cathode carbon blocks can replace a portion of coal[6]. The Al_2O_3 and SiO_2 in the waste cathode carbon block can be used as raw materials for cement production and enter the production process. Not all cement plants can utilize waste cathode carbon blocks, and cement plants with strict restrictions on production chemical or mineralogical composition cannot utilize high alkaline waste cathode carbon blocks. Usually, the amount of waste cathode carbon blocks added to cement is 0.2%. Aluminum metal particles are harmful in finished cement, so the added materials should be as uniform as possible and free of any aluminum metal particles. This method utilizes the carbonaceous material and alkali metal fluoride in the waste cathode carbon block, but mainly utilizes the waste refractory material in the spent potlining.

2.1.2 Steel industry

At present, in the steel industry, the cupola for melting iron requires metallurgical coke as fuel and fluorite as flux. Fluorite is a widely used auxiliary flux that can promote slag formation during the process of molten iron melting, facilitate the removal of non-metallic oxides in coking coal ash, and reduce the viscosity and liquid phase temperature of the slag. Therefore, it can improve the process conditions in the steel smelting process. Metallurgical coke and fluorite are both relatively expensive raw materials. The carbon contained in the waste cathode carbon block can be used as a fuel to replace metallurgical coke. The waste cathode carbon block also contains a considerable amount of fluorine, so fluorine salts and limestone can be mixed as additives to replace fluorite. The use of waste cathode carbon blocks as fuel and the reduction of slag viscosity in the steel industry has been reported[7]. The amount of waste cathode carbon blocks added per ton of pig iron is 5-25 kg, and the refractory material and electrolyte in the spent potlining are also dissolved in the slag.

The waste cathode carbon block was tested in a cupola, and the operation of the cupola was normal. The fluidity of the slag was greatly improved, and the sulfur and phosphorus contents were significantly reduced, resulting in high-quality gray cast iron. However, the waste cathode carbon block causes severe corrosion to the equipment lining, and its economic rationality needs to be discussed[8].

2.1.3 Mineral wool industry

Usually, mineral wool is made of coke and blast furnace slag, while glass wool is made of sand and coke. Preliminary experiments have been conducted in France and Norway on the use of spent potlining for mineral wool industry production, but the waste cathode carbon blocks in the spent potlining are only used as additives or substitutes for coke in the German mineral wool industry, and the addition amount is very small. The waste cathode carbon blocks in the spent potlining of a smelter in France were sent to the German mineral wool industry, and the waste refractory materials were sent to the cement industry[9].

2.1.4 Electric power industry

Scholars have conducted research on the mixed combustion technology of waste cathode carbon blocks and coal for power generation. In order not to cause environmental damage, a maximum of 8% of waste cathode carbon blocks can be added to the electric coal. At this time, the waste cathode carbon blocks do not have an impact on the scaling of the boiler wall. The combustion kinetics of coal with a small amount of waste cathode carbon blocks added are very similar to those of coal combustion alone. Givens[10] suggests that waste cathode carbon blocks can be used as fuel for coal-fired power plant boilers. No form of cyanide was detected in the discharged gas and in the bottom ash, but a certain amount of fluoride can be detected in the flue gas. When coal is used as fuel alone, the average fluoride content in the flue gas is 3.1 mg/m^3, while when coal with 2% waste cathode carbon block is added as fuel, the average fluoride content in the flue gas is 10.5 mg/m^3. Due to the high content of fluoride, the use of waste cathode carbon blocks alone for combustion and power generation treatment is not an ideal method, and therefore has not been widely applied in industry.

2.1.5 Landfill

More than 50% of the spent potlining is landfilled without any special treatment. The legislation varies from country to country, but currently the usual minimum requirement is to have dedicated storage sites or landfill pits with impermeable substrates[11]. In order to prevent fluoride poisoning in the drinking water of the early partially enclosed spent potlining landfill site, it is necessary to drill holes and extract groundwater under the landfill site for cleaning. Norway and Iceland conducted early stacking experiments in coastal basins. The report states that cyanide in the spent potlining is oxidized in the basin, and the leaching solution containing fluoride is not a toxic substance for seawater. Currently, Hydro is filling the spent potlining at a quarry on a small island in the Oslo Fjord, but the landfill is expected to be filled by 2022.

Although a large amount of experimental research has been conducted on waste cathode carbon blocks both domestically and internationally, and various wet and pyrometallurgical treatment technologies have been proposed, most primary aluminum plants do not have waste cathode carbon blocks for treatment due to treatment costs, technological maturity, and other reasons. Since 2018, various solid pollutants in the aluminum electrolysis industry have been included in the list of hazardous solid waste. Under the pressure of this policy, various aluminum factories have built several treatment production lines. But these treatment technologies all have drawbacks such as high investment, poor product quality, high operating costs, and the generation of new solid waste after treatment. Therefore, the development of a series of key technologies and equipment for the clean treatment and resource utilization of solid hazardous waste in China's aluminum electrolysis industry, as well as achieving the goal of treatment and resource utilization of solid hazardous waste in aluminum electrolysis, is an urgent challenge to be overcome, and it is also a technological bottleneck that must be solved for the sustainable development of China's aluminum industry.

At present, there are mainly two types of disposal methods for waste cathode carbon blocks in aluminum electrolysis at home and abroad: wet method and fire method. In terms of wet treatment, the main technology used by Rio Tinto Aluminum is the "LCL&L" technology, which is the low alkali leaching lime treatment spent potlining process[12]. Using water and dilute alkali solution to leach cyanide and fluoride from the spent potlining, cyanide is removed in a pressurized reactor, the concentrated solution is evaporated and reacts with lime, solidified and defluorinated, and the alkali solution is reused. But this process is complex and has high investment costs. In terms of fire treatment, the main process is to crush the waste cathode carbon block and then burn it by adding different additives (such as brown coal, limestone, etc.) to decompose cyanide and volatilize fluoride before landfilling. This method only achieved partial harmless treatment, resulting in resource waste. Aluminum Corporation of America adopts the Ausmelt pyrometallurgical process to treat spent potlining. Flux is added to the spent potlining and treated in an Ausmelt furnace at a treatment temperature of 1300 ℃. HF is recovered to generate aluminum fluoride, and the final product is a glassy slag that is difficult to further utilize.

The research on the waste lining of aluminum electrolysis cells in China started relatively late. In the late 1980s, research began on the treatment and recovery of waste cathode carbon blocks from aluminum electrolysis cells. Flotation methods were used to grind the waste cathode carbon blocks and add them together with water and flotation agents to the flotation cell. After multiple floatations, fluoride salts and carbon powder were obtained, but the leaching toxicity of the carbon powder still exceeded the standard. At the beginning of the 21st century, the technology of using grinding flotation acid leaching evaporation process to treat waste cathode carbon blocks was developed. However, due to the long process flow and high cost of wastewater treatment generated during the process, it cannot become an economic, reliable, and efficient treatment technology. The harmless treatment technology for aluminum electrolysis waste cell lining, developed in 2003, uses limestone and industrial waste PCA as additives to treat the waste cell lining. Through the technical route of burning carbon, oxidizing and decomposing cyanide, and converting fluoride, the harmless treatment of solid slag is achieved. Although this method can effectively solve the problem of pollution discharge from the waste lining of electrolytic cells, its process energy consumption is high and can only be treated as unusable solid waste residue, which has poor economic efficiency and has not been widely applied.

Although a large amount of experimental research has been conducted on the utilization of waste cathode carbon blocks both domestically and internationally, and various wet[13-16] and fire[17-21] treatment technologies have been proposed, these treatment technologies all have drawbacks such as high investment, poor product quality, high operating costs, and the generation of new solid waste after treatment. Therefore, there is currently no large-scale production line for the treatment of waste cathode carbon blocks put into operation.

To this end, the research and development of safe treatment technologies for solid/hazardous waste in the aluminum industry, such as aluminum dross and waste cathode carbon blocks, is the only way for the green and healthy development of China's aluminum industry.

2.2 Classification, composition, and phase

At present, the average life of aluminum reduction cells in China is more than 2600 d. In the production process for six or seven years, cathode carbon blocks and refractories will be impregnated with a large amount of fluoride salts, electrolytes and metal aluminum[22]. When the cell is overhauled, the weight of the lining material is nearly doubled. These waste lining materials, also known as Spent Pot-lining(SPL). Fig. 2-1 shows the SPL in the cell, which can be divided into spent cathode carbon block, waste silicon carbide side block and refractory material.

Fig. 2-1　SPL in the aluminum reduction cell

The lining material of the reduction cell is subject to double erosion of high temperature and molten fluoride salt for a long time. The fluoride salt and part of the air penetrate into the gaps of the lining material, causing the SPL to contain a large amount of fluoride and part of cyanide.

2.3 Characterization analysis

2.3.1 Composition of spent cathode carbon block

The cathode carbon block used in aluminum reduction cell is produced by electro-calcining anthracite, graphite crushed coal and coal bitumen by kneading, forming, roasting, graphitization and other processes. Most of the cathode carbon blocks are graphite cathode carbon block having a graphite content of 30% and 50%, wherein the graphitized carbon is about 60%-70%[23].

In addition to the graphitized carbon material, the spent cathode carbon block also contains 20%-30% fluoride salt, aluminum and alumina due to penetration and erosion by the electrolyte. Therefore, the graphite material and the fluoride and aluminum compounds in the spent cathode carbon block are all valuable resources. The spent cathode carbon block is shown in Fig. 2-2.

Fig. 2-2 Spent cathode carbon blocks yard

The primary aluminum plant produces approximately 8-10 kg of waste cathode carbon blocks for each ton of aluminum produced, and China will produce 250-350 kt of waste cathode carbon blocks for aluminum electrolysis cells every year. At present, the relatively standardized treatment method for large and medium-sized primary aluminum enterprises is to construct a dedicated disposal site for electrolytic cell waste lining for landfill treatment in accordance with the "Pollution Control Standards for Hazardous Waste Landfill". However, due to the high construction and maintenance costs, limited implementation, and incomplete governance, there has been no substantial physical or chemical treatment of the pollution source. More often, waste is simply landfilled or stored in the factory, or outsourced to a local hazardous waste treatment center for disposal, while treating 1 ton of hazardous solid waste requires a disposal fee ranging from 2000-4000 CNY. Due to lax regulation, the treatment of waste cell liners in some aluminum electrolysis enterprises is even more chaotic, with arbitrary disposal and exposure to sunlight and rain, resulting in frequent pollution incidents.

Waste cathode carbon blocks contain valuable carbonaceous materials, fluoride salts, and cryolite. But they are all mixtures with complex chemical composition, especially the electrolytic cell lining that is corroded and infiltrated by high-temperature electrolytes during long-term production. Therefore, the spent cathode carbon blocks generated after the overhaul of the reduction cell contain soluble F^- and harmful substances[24-26]. When these substances are stored outdoors and come into contact with humid air or rainwater, their soluble fluorine and cyanide will dissolve into the soil and groundwater, releasing harmful gases. If not handled properly, it will cause great harm to the local environment and society[27]. The generation of such a large amount of industrial hazardous waste is not only a great burden on the environment, but also a serious waste of precious resources. How to achieve clean and efficient treatment of toxic and harmful substances, and to recycle valuable substances as resources, is a technical challenge that urgently needs to be solved by Chinese aluminum industry scientists and technicians. As shown in Fig. 2-3, domestic and foreign scholars have conducted relevant exploration and research on the application of waste cathode carbon blocks in different fields.

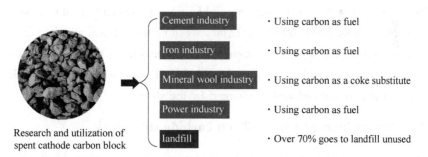

Fig. 2-3 Recycling and research status of spent cathode carbon blocks

The morphology of the spent cathode carbon block was analyzed by optical microscopy. A large amount of impurities were present inside the waste cathode carbon block. Further, the scanning electron microscope combined with the XRF analysis results (Table 2-1) shows that the spent cathode carbon block contains a large amount of elements such as F, Na, Al, Ca, Si, etc. in addition to C. Different forms of permeate can be found in the pores of the spent cathode carbon block.

Table 2-1 Chemical components of spent cathode carbon block

Element	C	F	Al	Na	Si	Ca	K	Fe	Others
Content(wt%)	64.93	12.94	6.38	7.85	0.74	1.22	0.61	0.39	4.94

The phase composition of the spent cathode carbon block was analyzed by X-ray diffraction. The results showed that the spent cathode carbon block contained a large amount of soluble fluoride such as NaF and Na_3AlF_6, (Table 2-2) and the total fluoride content was usually 20%–30%. In addition, a small amount of cyanide such as NaCN, $Na_4Fe(CN)_6$, and alumina and aluminum carbide are also present in the waste cathode carbon block. These fluorides and other solid reaction products that have penetrated into the gaps of the cathode carbon block have been tightly bonded to the carbon material. In addition, the carbon is almost graphitized, and it will slip when it is broken by the jaw type, which causes great difficulty in crushing.

Table 2-2 Phase components of spent cathode carbon block

Sample	Phase components	Chemical formula	Content(wt%)
Spent cathode carbon block	Carbon	C	60–70
	Sodium fluoride	NaF	10–15
	Fluorite	CaF_2	5–10
	Cryolite	Na_3AlF_6	8–12

After leaching toxicity analysis, the soluble fluoride content in the waste cathode carbon block is about 3000–6000 mg/L, far exceeding the national standard of 100 mg/L for hazardous waste. The cyanide content is 10–20 mg/L. In the past, the accident of cyanide poisoning occurred when the electrolytic cell was removed. It is a dangerous solid waste that is monitored and discharged by the state.

2.3.2 Composition of waste silicon carbide side block

With the increase of the capacity of the reduction cell, the carbon block on the side cannot meet the heat balance requirement. The silicon carbide combined with the silicon nitride materials are widely used as side blocks of aluminum reduction cells because of their excellent thermal conductivity, thermal shock resistance, insulation and resistance to high temperature molten electrolyte erosion[28].

However, in the production process of more than 2000 d, Si_3N_4 and SiC in silicon carbide bricks were continuously oxidized to SiO_2[29]. The waste silicon carbide side block is shown in Fig. 2-4.

Fig. 2-4 Typical waste silicon carbide side block
from the aluminum reduction industry

The lining of the side of the reduction cell is gradually destroyed as the electrolysis process continues. The side silicon carbide bricks are in direct contact with the high temperature molten electrolyte. The silicon carbide brick becomes loose and brittle. Some of the fluoride salt also penetrates into the pores of the silicon carbide brick under the action of carbon oxides and air. The amount of silicon carbide side block of a 400 kA cell is 7 t, worth about 80000 CNY. Such high-energy materials can be widely used if impurities such as fluoride salts are removed.

The phase composition of the s waste silicon carbide side block was analyzed by X-ray diffraction. The results showed that the main components of the waste side block were SiC, Si_3N_4, NaF, SiO_2, SiF_4, and Na_3AlF_6 (Table 2-3). The gap of the side lining material is infiltrated by the electrolyte, and the hard and dense waste side block is formed, and it is also difficult to be broken and ground. There are differences in the components of the waste side blocks in different parts, and the main harmful substances are fluoride salts.

Table 2-3 Chemical components of waste silicon carbide side blocks

Composition	SiC	Si_3N_4	SiO_2	SiF_4	NaF	Na_3AlF_6	Others
Content(wt%)	76.72	14.37	1.73	0.74	2.59	2.08	1.77

2.3.3 Composition of waste refractory material

The bottom refractory material of the reduction cell is composed of dry refractory material and refractory brick. The dry refractory material is a key component to prevent electrolyte from infiltrating into the thermal insulation material. After 6-7 a of service in the reduction cell, the dry refractory material reacts with the infiltrated electrolyte to form a dense glassy material of 5-15 mm[30]. This vitreous material blocks further penetration of the electrolyte and protects the underlying insulation material from corrosion.

The phase and chemical composition of the waste refractory material were analyzed by X-ray diffraction and XRF(Table 2-4), and the results showed that the main components of the waste refractory material were SiO_2, sodium alumina, and fluoride salt. The content of F and Na in the refractory material gradually increases, and the main component is fluoride salt. It is proved that the electrolyte has rapidly penetrated into the refractory material at this time, and the reduction cell needs to be overhauled.

Table 2-4 Chemical composition of waste refractory material

Element	Si	Al	Fe	Ca	F	Na	K	O
Content(wt/%)	6.71	29.02	1.20	2.51	28.15	17.11	2.08	11.94

2.4 Toxic substance footprint

Spent pot lining (SPL) is the hazardous solid wastes generated by aluminum reduction cells. The distribution status of fluorides and cyanide in a 350 kA cell operated for 2396 d was analyzed and the footprint and corrosion mechanism of the harmful substances in SPL were studied. The fluorides are mainly concentrated in the cathode carbon block and the layer of dry barrier under the cathodes which is closely related to permeability of the cathodes and dry barrier the fluorides penetrate in. Cyanide has a low concentration in the cell center and a high concentration in the sidewall, which is related to the air amount entering into the areas in the cells.

In the aluminum reduction process alumina is dissolved in cryolite in electrolytic cells called pots that consist of steel shells lined with carbon and refractory materials. The inner portion of the lining serves as the cathode which contains the molten electrolyte. The carbon cathodes are composed of prefabricated carbon blocks rammed together and seamed by the carbon paste. The sidewalls of the lining typically are formed with prefabricated carbon blocks and silicon nitride bonded carbide combined blocks. Insulation packages for a cell are mostly of dry barrier, refractory bricks and insulating materials.

The high temperature molten salt system is immersed in the reduction cells and directly contact with the lining materials for a long time. The electrolyte will penetrate into outsides bringing about a large amount of fluorides existing in the cathodes and lining. At the same time, air will infiltrate into the electrolytic cell during operation and react with the carbonaceous materials and bath to form highly toxic cyanide[31-33].

The toxic soluble fluoride and cyanide in the SPL may be leached and discharged into environment such as soil, rivers and groundwater in humid air or during rainy weather. Harmful gases emission will also take place during this process. Study results have shown that the lethal dose of cyanide is only 0.001-0.002 g/L. The SPL seriously threatens the living environment of surrounding humans, livestock and vegetation[34].

About 30-40 kg of SPL is discharged for production of per ton of aluminum. It is shown that the raw aluminum output in 2017 was 32.27 Mt in China bringing about discharge of 0.9 Mt of SPL. If so much hazardous materials are not effectively treated and become harmless the serious consequences will come out. As the result SPL has been classified now as the dangerous solid wastes by many countries and regions.

With the rapid development of the world aluminum industry, the electrolysis process parameters and lining design have been optimized. The total dumped SPL is also quickly increasing. The treatments of SPL and environment pollution elimination have become more and more urgent.

A lot of work has been done by the scholars and engineers on the harmless treatment and resource utilization of SPL[35]. Shi et al.[36] separated the cryolite from the SPL by two-step acid-caustic leaching. Saterlay et al.[16] processed SPL by using ultrasound, which has a faster leaching speed and leaching rate than conventional leaching. Cyanide is destroyed by oxidation of hydrogen peroxide under the ultrasonic surroundings. Caesar Company of the United States invented a method for treating SPL at high temperature[37]. Adequate water is introduced to hydrolyze fluoride and cyanide at 1100-1350 ℃ to form HF gas and gaseous NaF. The residue contains sodium oxide and alumina. Reynolds[38] developed a detoxification process where a blend of SPL, limestone and an anti-agglomeration agent are calcined in a rotary kiln using natural gas as the heat source. The concept is that CaO will convert $NaF-AlF_3$ to CaF_2 which is much less soluble. The cyanides are also destroyed by the high temperature process. Chalco SPL[39] treatment is a pyro-process at the temperatures of 900-1050 ℃, which also destroys all cyanides rapidly. The detoxified solid residue is recycled to cement production and the fluorides to the alumina scrubbers.

Not so much study results on the formation mechanism and distribution of harmful substances in SPL is presented. In 1985, Peterson et al. conducted a study distribution of total fluorides and cyanides within cell linings for multiple cell technologies (65-225 kA), followed by laboratory experiments to confirm the suspected mechanism of cyanide formation. However, the low current intensity cell is gradually replaced by high current intensity cell. The 350 kA cell is used by most companies now. The distribution of fluoride and cyanide in the lining material has also changed due to the different operating parameter and cell lining design. Silveira characterized the leaching extent of cyanides and fluorides from SPL as a function of the number of years when the lining materials were presented in an operating cell[40]. However, it seems that there is no clear correlation between the hazardous materials and the cell life. It is found during our processing SPL that the compositions of SPL are highly variable (e. g. cyanide, fluorides and metals), but the components of greatest concern environmentally are also cyanide and soluble fluoride salts. The SPL has different components in different cells and smelters, which need to be treated in different ways. And the products obtained after treatment are also different.

Studying the behavior of toxic substances in different types of SPL is of great significance for the further targeted treatment. The study of the diffusion mechanism of hazardous materials in the lining material can also optimize the lining design of the aluminum reduction cell. The purpose of this study is to analyze the content distribution of harmful substances in SPL at the typical locations of cells and to study their footprint and corrosion mechanism. It is our purpose to provide basis for the green processing and resource utilization of different parts of SPL.

The samples used in this experiment were taken from a 350 kA cell with a life of 2396 d in an aluminum smelter in the western China including the waste cathode carbon blocks, dry barrier, refractory and insulation bricks etc (Fig. 2-5).

Firstly the samples were taken at the different sampling points in the cell by a pneumatic pick drilling. Then the bath agglomerates stuck on the sample surface were cleaned up. After preparation the samples were put in the sample bags and placed in a constant temperature 80 ℃ desiccator for 24 h. Finally the dried sample was placed in a sample grounding machine for particle size range of −100 mesh for subsequent chemical and mineral analysis and research tests.

The cyanide content was measured by using the silver nitrate titration method (the minimum detection concentration of cyanide is 0.25 mg/L). This method is to use a 0.01 mol/L silver nitrate standard solution to titrate the SPL leachate to which the indicator is added. The cyanide ion reacts with silver nitrate to form $Ag(CN)_2^-$, and an excess of silver ions reacts with the indicator to change the color from yellow to orange-red. The concentration of cyanide can be calculated based on the amount of silver nitrate

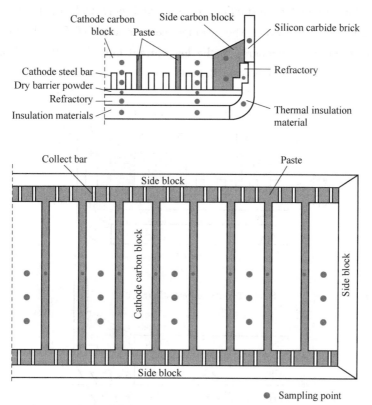

Fig. 2-5 Distribution of sampling points in the lining material of the electrolytic cell

at the titration end.

The leachable fluorine content in the samples is analysed by the ion selective electrode method in GB/T 15555.11(the minimum detection concentration is 0.05 mg/L). And the instrument of the ion activity meter(PXSJ-216, Shanghai INESA Scientific instrument Co., Ltd, China) is used for leachable fluorides analusis.

It is found out from this study that the distribution of hazardous materials in the SPL is mainly influenced by the design of the lining, the cell life and the process parameters.

Unlike the cell with low current intensity in the past, the life of the high-current cell with improved structure can reach more than 2000 d. More fluoride was accumulated into the lining material as the life of the cell increases. There is more or less fluoride diffusion in each part of the lining material in the cell after 2396 d. The average concentration of soluble fluoride in the cathode carbon block is 3422 mg/L, 3207 mg/L for the paste between the cathode carbon block, 3705 mg/L for the side carbon block, 1702 mg/L for the silicon carbide side block, 3077 mg/L for the side refractory material, the dry barrier power is 8315 mg/L, the bottom refractory material is 2696 mg/L,

1296 mg/L for the bottom insulation material and only 691 mg/L for the thermal insulating material (Fig. 2-6). The average concentration of fluoride in the cell is increased by 2000 mg/L compared to the low current intensity cell, which is related to the age of the cell and the amount of fluoride used.

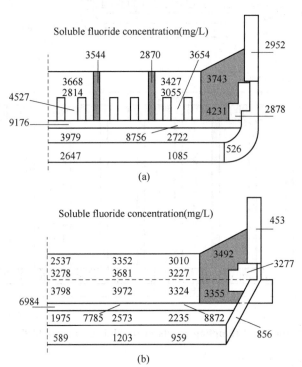

Fig. 2-6　Distribution of fluoride in longitudinal (a) and lateral (b) of 350 kA cell

The results show that there is a large amount of fluorides in the spent cathode carbon block which are mainly in the form of sodium fluoride, calcium fluoride and cryolite (Fig. 2-7(a)). In the new cathode carbon block there exsits only carbonaceous material (Fig. 2-7(b)).

Most of the fluoride is concentrated in the carbon block and the dry barrier layer and the highest concentration of soluble fluoride in the lining appears below the cathode carbon block.

From the top to the bottom of the cathode carbon block the fluoride is evenly distributed, and there is no obvious concentration gradient. The fluoride concentration is slightly increased at the bottom of the cathode. But the concentration of soluble fluoride rises sharply in the dry barrier layer and reaches the peak content in the entire cell. This

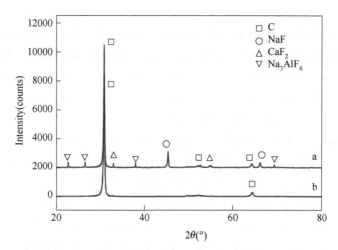

Fig. 2-7　XRD pattern of spent cathode carbon block(a) and new cathode carbon block(b)

means that a large amount of fluoride is enriched in this area. However, after passing through the dry barrier layer the concentration of soluble fluoride in the refractory material drops dramatically, which is even much lower than that in the cathode carbon block. Most of the fluoride is isolated in the dry barrier layer and it is difficult to penetrate further into the bottom lining materials. Only small amount of the fluoride can be detected in the bottom insulation material, which is much lower than that of other locations(Fig. 2-8).

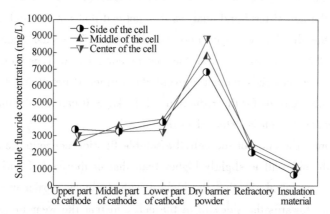

Fig. 2-8　Fluoride concentration gradient in the vertical direction of the cell lining

The cathode carbon blocks are the most important structural part of the aluminum reduction cell, which plays an important role in conducting current and undertaking high temperature molten salt. The cathode carbon blocks have been in direct contact with the metal and bath for several years and the fluoride in the electrolyte can gradually diffuse into this region. As the electrochemical reaction proceeds, the cathode carbon block is worn and cracked under the actions of erosion and stress in the molten aluminum metal. Moreover, fluoride can further penetrate into the lining material due to the presence of these cracks on the cathode surface. As time goes by, the fluoride content in the cathode carbon block has accumulated to a higher level.

Under the bottom of the cathode carbon block is a dry barrier layer in which main components are Al_2O_3 and SiO_2. When the fluorides diffuse to under the carbon block, these fluorides can react with the dry barrier materials in this area (Eq. (2-1), Eq. (2-2)), and the main reaction products are nepheline and albite[41]. Nepheline will exist in the form of a crystalline, and albite will form a sticky glass layer at the reaction site which will prevent further penetration of fluoride into the next layer. Moreover, the temperature at this location is also suitable for solid phase deposition and columnar crystal formation. Consequently, the fluoride gradually accumulates under the bottom of the cathode carbon block, and the fluoride concentration reaches the peak of the entire electrolytic cell here.

$$6NaF(l) + 3SiO_2 \cdot 2Al_2O_3(s) = 3NaAlSiO_4(s) + Na_3AlF_6(s) \quad (2\text{-}1)$$

$$6NaF(l) + 9SiO_2 \cdot 2Al_2O_3(s) = 3NaAlSi_3O_8(s) + Na_3AlF_6(s) \quad (2\text{-}2)$$

Further penetration of the fluoride can be hindered by the presence of the dry barrier layer due to the reactions so that the concentration of fluoride in the refractory brick and insulation material below the barrier layer is much lower (Fig. 2-9(a)).

However in the actual tests and analysis it is noticed that not so much of the fluorides can penetrate the dry barrier layer into the refractory. When too many fluorides are infiltrated into the refractory material the severe corrosion will happen in this area. Corroded refractory materials cannot prevent the further penetration of the fluorides, which is one of the reasons for the reduction cell leakage. It can also be found that a very high fluoride concentration is detected in this area.

In the horizontal direction of the cell, the soluble fluoride concentration of the carbon block besides the sidewall is slightly higher than that of the central carbon block. It is shown by Tabereaux et al.[42] that the highest cathode block erosion was often located under the anodes towards the sidewall of the cell and that the wear under some anodes was deeper than under others. It is also found that there is almost no erosion in the center of the cell. This uneven form of cathode erosion leads to the most common erosion pattern observed in modern high-amperage prebaked anode cells called the W or WW wear pattern (Fig. 2-9(b)). The formation of the W wear pattern is related to the current

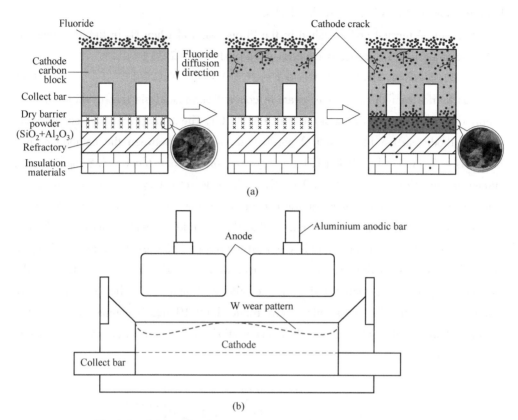

Fig. 2-9 Schematic diagram of fluoride diffusion into the lining material (a) and W shape wear pattern (b)

distribution and flow field motion in the cell. This corrosion pattern is more pronounced in high current intensity cells. In the areas where the corrosion pit is deep the higher concentrations of fluorides will be detected than that of other positions. Potholes can be easily found in the areas with a high degree of corrosion. This is helpful for the deposition and further diffusion of fluorides in this area. It can be supposed that the distribution of fluorides in the lining materials of the aluminum reduction cells is probably related to the current intensity and the flow status of the molten metal.

For the sidewall, the concentration of soluble fluoride in the side carbon block is higher than in the side block of silicon carbide the reason of which is possibly related to the side carbon block formation and baking process of the cells. The side wall includes the silicon carbide blocks and side carbon blocks. The carbon blocks are mainly from the ramming paste and baked during the baking process of the cells when lots of fluorides are absorbed into the ramming paste so that the side carbon blocks contain much more fluorides than in the silicon carbide blocks.

NaCN can be formed and enriched in the lining material when N_2 in the air meets the sodium-containing carbonaceous material in the cell (Eq. (2-3)).

$$1.5N_2(g) + 3Na(s) + 3C(s) = 3NaCN(l) \quad (2\text{-}3)$$

It was found from the results of cyanide concentration analysis that cyanide was not evenly distributed in the lining of the cell (Fig. 2-10). The average cyanide concentration in the cathode carbon block is 5.2 mg/L and the paste between the cathode carbon blocks is 4.5 mg/L while it is 61.1 mg/L in the side carbon block, 73.8 mg/L in the silicon carbide side block and 59.9 mg/L in the side refractory material, which means side wall contains much higher cyanide concentration than in the cathode blocks on the cell bottom. It is because the same reason that the cyanide concentrations in dry barrier layer, bottom refractory material and bottom insulation material are 7.8 mg/L, 5.2 mg/L and 3.7 mg/L respectively. The cyanide concentration is as high as 73.3 mg/L in the side thermal insulating material like in the side wall materials. The concentration of cyanide in the cathode carbon block was reduced by 10 mg/L compared to the low current intensity cell, but the cyanide concentration in the side liner was significantly increased by 50 mg/L.

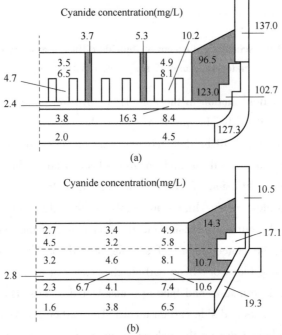

Fig. 2-10 Distribution of cyanide in longitudinal (a) and lateral (b) of 350 kA cell

A very low level of cyanide is detected in the central region of the cell. However, the closer to the side wall of the cell, the higher the cyanide concentration is in this sample cell. The concentration of cyanide in the side lining material can be several times higher than that in the cell center. It is found as well that cyanide also appeared in non-carbonaceous materials. This means that cyanide is not fixed in the place where it is formed, but can be diffused and transferred into the lining material at other locations in the cell. It is found that the cyanide concentration in the silicon carbide side block and carbon block near the aluminum suction positions is also much higher than that in the other areas (Fig. 2-11).

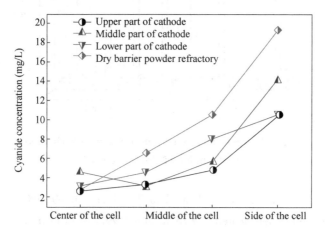

Fig. 2-11 The cyanide concentration gradient in the horizontal direction of the lining

The cyanide is produced by the reaction of the carbonaceous material with nitrogen in the air. This means that a high concentration of cyanide can be easily detected in a region where is in contact with air and exist carbonaceous materials. Air can penetrate into the cell by the portholes in the side walls during aluminum reduction. Most air is infiltrated around the side walls of the cells and difficultly enters the cell center. Therefore, high levels of cyanide appear in the side lining material of the reduction cells. In the same way, the cyanide concentration in the cell center is much lower due to lack of air.

The aluminum suction position is a special place for cyanide formation. The operations is carried out every day at this position, such as salvaging the carbon residue, measuring the levels of bath and aluminum metal, observing the operation state of the cells and extinguishing the anode effects and so on. Frequent breaking of the crust and agitation of the bath and metal will cause a large amount of air to flow into this area, which creates good conditions for the formation of cyanide. That is why there is a high concentration of cyanide in the silicon carbide side block and the cathode carbon block near the suction position. The cyanide content in the lower side wall area is very low because it is immersed in the bath and metal and there is less chance to contact with air. So the cyanide content in the lining is closely related to if air can enter into this part easily (Fig. 2-12).

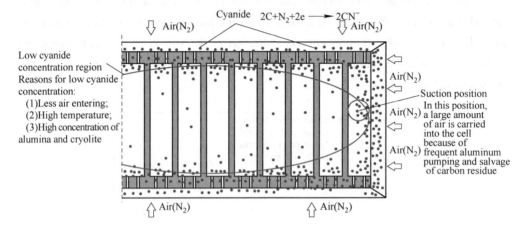

Fig. 2-12 Schematic diagram of cyanide distribution in the lining of electrolytic cell

According to the literature, when heated to 773 K, about 99.5% of cyanide in SPL was decomposed, and when heated to 873 K, about 99.8% of cyanide was decomposed. When heated to above 973 K, cyanide completely disappears. We found that cyanide is prone to decomposition in an aerobic atmosphere. This means that although the infiltration of air in the cell causes the formation of cyanide, cyanide can also be decomposed in the presence of oxygen in the high temperature region. Therefore, a higher concentration of cyanide can be found in the side regions where the temperature is slightly lower and the air is more easily penetrated. However, cyanide was hardly found in the central portion of the cell where the temperature was high.

Besides NaCN can be unstable when Na_3AlF_6, Al_2O_3 and Na are simultaneously present in the reduction cell (1223 K) based on Eq. (2-4) and Eq. (2-5).

$$1.5Na_3AlF_6(l) + 1.5NaCN(l) + 3Na(s) \Longleftrightarrow 1.5AlN(s) + 9NaF(l) + 1.5C(s)$$
$$\Delta G^{\ominus}_{1223 \text{ K}} = -255.0 \text{ kJ} \tag{2-4}$$

$$3Al_2O_3(s) + 1.5NaCN(l) + 3Na(s) \Longleftrightarrow 4.5NaAlO_2(s) + 1.5AlN(s) + 1.5C(s)$$
$$\Delta G^{\ominus}_{1223 \text{ K}} = -214.8 \text{ kJ} \tag{2-5}$$

The alumina feeders are located on the central axis of the electrolytic cell, so that the central portion of the cell is more enriched in alumina, which is prone to carry out this decomposition reaction so that there is a lower cyanide concentration in this region while a higher cyanide distribution on the cell side where is less Al_2O_3 concentration.

A cyanide distribution pattern is shown as above mentioned in which it is lower in the cell center and it becomes higher in the surrounding lining of the cells.

Fluoride and cyanide in the SPL are the main factors affecting the environment, and the leachate from the long-term stacking SPL will contaminate the soil and groundwater nearby. The 350 kA cell has been further improved in operating parameters and lining structure design compared to the low current intensity cell, with an average cell age of over 2000 d. The distribution characteristics of soluble fluoride and cyanide in the lining materials are also somewhat different.

The average concentration of fluoride in high current intensity electrolytic cells is higher than that in a low current intensity electrolytic cells, which is related to the operating process system of the electrolytic cells. Fluorides are usually concentrated in the cathode carbon block and dry barrier layer. The highest concentration of fluoride appears below the cathode carbon block, where the reaction between fluorides and dry barrier will happen. The vitreous material formed in this reaction isolates most of the fluorides into this area and prevent them to further diffuse.

The concentration of cyanide in the cathode carbon block was reduced by 10 mg/L compared to the low current intensity cell, but the cyanide concentration in the side liner was significantly increased by 50 mg/L. This is related to the better tightness of the cell and the higher two levels. However, frequent air infiltration at the operating end also results in a high content of cyanide in the side liner. However, a large amount of cyanide was found in the side lining because of frequent air infiltration at the operating end. The cyanide content in the various parts of the cell lining is closely related to the amount of air that can enter into this part. Cyanide is mostly distributed near the side wall of the cell. It is not static, it will diffuse and transfer in the electrolytic cell.

The process concept for the classification and treatment of the hazardous materials in SPL could be put forward based on the study results of the harmful substances footprint in SPL.

2.5 Pyrometallurgy treatment technology

2.5.1 Experimental study on fluorine separation

The spent cathode carbon block (SCCB) from aluminum reduction cells is considered to be a hazardous material due to containing a large amount of soluble fluoride salts and toxic cyanides. The serious environment pollution and ecological destruction will be caused by being openpiled or buried of SCCB. A joint temperature-vacuum controlling process for treating SCCB is proposed. The thermodynamic analysis by FactSage 7.0, series of experimental investigation, and characteristics tests on the product materials by SEM and XRD were carried out. The factors including temperature, vacuum pressure and retention time that influence the SCCB's detoxification effects were comprehensively investigated and discussed in detail. It was found out that the soluble fluoride content can be reduced to a level of 3.5 mg/L while cyanide is completely decomposed. A fixed carbon content of 97.89% in the treated materials is obtained under the optimum temperature and vacuum controlling conditions.

The primary aluminum in the world is mainly produced by the cryolite-alumina molten salt electrolysis process at present. The primary aluminum metal is reduced from the molten mixture of alumina, cryolite and the other fluoride salts in the reduction cell. In the aluminum smelting process the cathode carbon blocks serve as the structural part of the aluminum reduction cells and also play functions of conducting current and bearing the high temperature molten mixture.

However, the cathode carbon block would gradually wear under continuous chemical corrosion and physical erosion in the high temperature molten metal and salt. The molten mixture would penetrate into the cathode carbon block causing cathodes to expand[43].

The life of a reduction cell is generally 4-7 a. The cell needs to be overhauled when it does not meet normal production needs. A large amount of spent lining materials including about 1/3 of spent cathode carbon blocks would be produced. It was reported that 24-30 kg SCCB is discharged for producing 1000 kg primary aluminum and about 1.88 Mt of spent cathode carbon blocks are produced around the world every year[44].

The spent cathode carbon blocks (SCCB) contain a certain amount of toxic and harmful substances including fluoride salts and cyanide except the carbonaceous material. According to detection results the soluble fluoride salts in SCCB is at the level of 2000-4000 mg/L and much higher than the safe discharge standard of 100 mg/L. And the 10-20 mg/L of cyanide existing in SCCB is also higher than the safe discharge standard of 5 mg/L.

If the SCCB are not handled properly, for instance just openpiled or buried, the soluble fluorides and cyanide could dissolve and diffuse into the soil and groundwater and would cause very serious damage to ecological environment as opened to rain or moist air[45-46]. It was reported that the dental fluorosis for the residents would be caused

2.5 Pyrometallurgy treatment technology

by fluoride pollution and is 21.84% greater than the normal incidence of only 1.52%. Consequently, the SCCB is considered as hazardous solid waste in many countries because of their toxicity, leachability and reactivity with water.

A lot of research work on how to realize clean and efficient detoxification of SCCB had been done in the world in the past years. The methods for treating SCCB can be generally divided into hydrometallurgy and pyrometallurgy processes. The hydrometallurgical treatment is mianly to extract the fluoride from the SCCB in a solution environment. The traditional pyrometallurgical treatment is mainly to burn out the carbonaceous material at high temperature, during which the fluorides convert to discharge as the flue gas in the form of HF and the cyanide is decomposed as well.

For pyrometallurgical methods, Caesar Company in the United States invents a method for treating SCCB at high temperature. Adequate water is introduced at 1100−1350 ℃ to hydrolyze fluoride and cyanide, during which HF gas and gaseous NaF and residues containing sodium oxide and alumina are produced. This method does turn SCCB into harmless waste, but the disposal of these wastes has become a new problem. Similarly, Chalco's spent potlining was treated by a pyro-process at temperature of 900−1050 ℃, in which all cyanides were decomposed rapidly, the detoxified solid residue was recycled to cement production and the volatile fluorides were removed into the alumina scruber. Similarly, there is still the problem of nowhere to place the tailings, and the carbonaceous material in the SCCB is also wasted. To date, there is still no perfect way to handle SCCB in the aluminum reduction industry. In the course of the research, some researchers found that the application of external enhancement to detoxification of SCCB can achieve better results. Saterlay et al. destroyed and removed cyanide and cryolite in the SCCB by using ultrasound method. It has a faster leaching rate than the conventional leaching process. Cyanide is destroyed by oxidation of hydrogen peroxide produced by ultrasonication. Further, In Nan's manuscript, the new vacuum distillation method(VDM) was recommended to pretreat SCCB below 1000 ℃ to separate volatile fluorides and graphite. The products, leachable fluorides and carbon enriched residue, are expected to be reused in metallurgical industries for cost-saving. The way of applying external reinforcement is indeed superior to traditional processing methods.

Both the hydrometallurgical and pyrometallurgical methods have their own shortcomings. The hydrometallurgical treatment process generates a large amount of harmful gases such as HF and easily causes the serious corrosion of the equipment. The treated spent cathode blocks still contain a high soluble fluoride content(still considered as hazardous material). The high graphitization carbonaceous materials are used as fuel in the traditional pyrometallurgical process, which is a waste of carbon resource, and the products will be the useless waste residues.

A joint temperature-vacuum controlling process for treating SCCB is proposed as shown in Fig. 2-13. Briefly, the method controls a certain temperature and pressure to remove harmful substances during the heat treatment of the SCCB. The thermodynamic analysis was first carried out for the optimum removal conditions of the harmful substances in the SCCB.

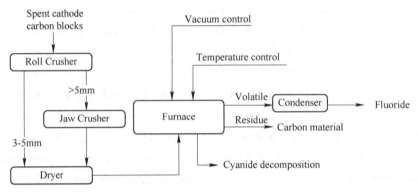

Fig. 2-13 Joint temperature-vacuum controlling process for treating SCCB

The SCCB taken from an overhauled aluminum reduction cell which had been in service for 2163 d in a smelter in western China was used in this study. The spent potlining is often the mixture of SCCB, spent side SiC block and spent refractory material so that the SCCB should be separated from the mixture and the electrolyte agglomerate and the refractory material adherent on SCCB should be removed. The raw SCCB was then dried in an oven at temperature 80 ℃ for 24 h to eliminate the moisture and was broken down by using a jaw crusher and then sieved to a particle size range of 3-5 mm, which was used as the raw material for subsequent experiments and characterization tests.

Table 2-5 lists the physical and chemical properties of the SCCB. As show in this table there are a high level of soluble fluoride content of 2342 mg/L and amount of cyanide of 10.2 mg/L.

Table 2-5 The physical and chemical properties of SCCB

No.	Physical properties		No.	Chemical composition(wt%)		Melting point(℃)
1	Particle size distribution(mm)	3-5	5	Fixed carbon content	74.08	—
2	Density(kg/m^3)	1.57	6	NaF	9.72	993
3	Soluble fluoride content(mg/L)	2342	7	Na_3AlF_6	4.86	1009
4	Cyanide content(mg/L)	10.2	8	CaF_2	2.43	1423

2.5 Pyrometallurgy treatment technology

The experiments were carried out in a pilot-level electric heating furnace consisted of high temperautre reactor, condenser and vacuum pump, as shown in Fig. 2-14.

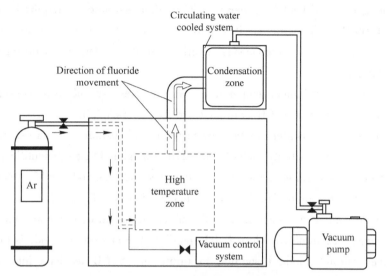

Fig. 2-14 Pilot-level electric heating furnace for treating SCCB

A vacuum control zone (including gas source Ar, vacuum pump, pressure control system), a high temperature reaction zone (maximum temperature of 1800 ℃) and a condenser (including circulating water cooling system) are set in the reactor. The vacuum pressure in the reaction zone can be adjusted at the high temperatures. The SCCB is treated in the high temperature zone, and the volatilized fluoride flue gas is condensed and collected in the condenser. The size of the heating zone is 300 mm×300 mm×300 mm, and the refractory material can withstand 1800 ℃, which can meet the requirements of the experiment. Two layers of stainless steel separators of different heights are placed inside the condenser to prolong the movement path of the volatilized fluoride, which is conducive to the condensation of the fluoride salt.

For each single test, we put (30±0.001) g of SCCB into a graphite crucible and placed in high temperature reaction zone. The detailed experimental parameters are set as shown in Table 2-6. The heating rate was 8 K/min at the temperatures lower than 1200 ℃ and 5 K/min when temperature was higher than 1200 ℃.

Table 2-6 Scheme for detoxification experiments of SCCB

Temperature(℃)	Vacuum pressure(Pa)	Retention time(h)
900-1700(interval 100)	3000(fixed)	3(fixed)
1500(fixed)	100000,6000,3000,1500,1000,500	3(fixed)
1500(fixed)	3000(fixed)	1,1.5,2,2.5,3

The treatment system was vacuumed before experiment start. When the system vacuum reaches the required level(such as 3000 Pa), the temperature was raised up according to the heating rate and kept for a period of time after reaching the required temperature (such as 1600 ℃). In particular it should be ensured that the argon atmosphere was maintained to protect the carbonaceous material from the oxidation risk during the entire experimental process.

The separation conditions (temperature, vacuum pressure) of toxic substances and carbonaceous materials in SCCB were analysed by FactSage7.0.

The soluble fluoride content in the raw material and residual in this experiment was measured by using ion activity meter(PXSJ-216, Shanghai INESA Scientific Instrument Co., Ltd, China), according to the ion selective electrode method(CETC, 1996)(GB/T 15555.11—1995, the minimum detection concentration is 0.05 mg/L).

The cyanide content was measured by the silver nitrate titration method(the minimum detection concentration of cyanide is 0.025 mg/L). The analysis methode is to use 0.01 mol/L silver nitrate standard solution to titrate the SCCB leachate into which the indicator is added. The CN^- reacts with silver nitrate to form complex, but an excess silver ions will react with the indicator to cause color changes. The concentration of cyanide is calculated based on the added amount of silver nitrate after titration.

The phase component of raw sample and residual is determined by X-ray diffraction analysis by using a PW 1710 diffractometer (XRD, Philips-Netherlands) and the morphology of the powders was tested using JSM-7500 F scanning electron microscope (SEM, JOEL-Japan).

The SCCB sample used in this study contains about 10.2 mg/L of cyanide, which mainly exists in the form of NaCN. NaCN could react with oxygen and can be converted to N_2, CO_2, NO, NO_2, and Na_2O. Table 2-7 lists three possible reactions at atmosphere of O_2 and the corresponding Gibbs functions.

Table 2-7 Cyanide decomposition reaction

Reaction	Gibbs function	No.
$2NaCN+2.5O_2 = 2CO_2+N_2+Na_2O$	$\Delta G_T = -728879+77.34T$	Eq. (2-6)
$2NaCN+4.5O_2 = 2CO_2+2NO_2+Na_2O$	$\Delta G_T = -638175+51.68T$	Eq. (2-7)
$2NaCN+3.5O_2 = 2CO_2+2NO+Na_2O$	$\Delta G_T = -697281+203.84T$	Eq. (2-8)

Fig. 2-15 shows the Gibbs free energy variation with temperature for cyanide decomposition reaction.

It can be seen in Fig. 2-15 that the trend of cyanide decomposition reaction is very sufficient. As shown in literature about 99.5% of cyanide in SPL decomposes when

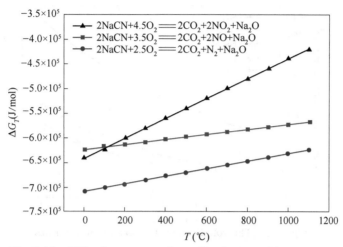

Fig. 2-15　Gibbs free energy of cyanide decomposition reaction

heated to 500 ℃ and about 99.8% of cyanide decomposes when heated to 600 ℃. Cyanide completely disappears as the temperature above 700 ℃. It can be expected that the cyanide in the SCCB can be effectively decomposed at the high temperatures.

By the joint temperature-vacuum controlling process different from traditional pyrometallurgical treatment the fluorides in SCCB can be effectively removed based on the saturated vapor pressure of each fluoride salt and the carbonaceous material can be recovered simulataneously.

The volatilization temperatures of NaF, AlF_3, Na_3AlF_6, LiF, KF and CaF_2 were firstly calculated under different conditions by the equilibar module of the thermodynamic software FactSage7.0. The vacuum pressure was selected to be 1 Pa, 10 Pa, 100 Pa, 1000 Pa, 10000 Pa, 100000 Pa, respectively.

Fig. 2-16 shows the thermodynamic analysis results of volatilization for the different fluorides. It can be found out based on the thermodynamic analysis that there is a similar volatilization trend for the different fluorides. The volatilization temperature of the fluorides decrease with the vacuum pressure drops, which means that fluorides can easily volatilize with lower vacuum pressure at the lower temperatures. In the other aspect, too low vacuum pressure will lead to a decrease of oxygen concentration in the system, then further affecting the cyanide decomposition reactions Eq. (2-6)–Eq. (2-8) in Table 2-7. It can be expected that a better fluoride and cyanide removal effect can be obtained under the lower temperatures and appropriate vacuum pressure.

The vacuum pressure of the system should be low enough for the fluoride removal as much as possible. However, the trace oxygen should be maintained for the destruction of cyanide as well. Therefore, the ideal vacuum pressure should be between 100 Pa and

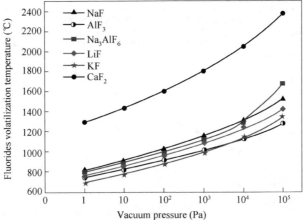

Fig. 2-16 The volatilization temperature of fluorides under different vacuum pressures

10000 Pa. It should be noted that the volatilization temperature of calcium fluoride (CaF_2) is much higher than that of other fluorides as shown in Fig. 2-16. Calcium fluoride is very stable and insoluble in water so that it does not contribute to the soluble fluoride content in the SCCB. It is concluded that the thermodynamic analysis indicates that almost all of fluorides (except CaF_2) in SCCB can theoretically be removed below 1500 ℃ under the different vacuum pressure.

A conclusion can be drawn from the above thermodynamic analysis results that the cyanide is easily destroyed during the high temperature treatment, while the temperature and vacuum pressure are two important factors affecting the fluoride removal in SCCB. The higher the treatment temperature and the lower vacuum pressure is and the better the fluoride removal from the SCCB.

In order to study the vacuum pressure effect on the removal of fluorides and cyanides from SCCB the heat treatment experiments under the maximum temperature of 1500 ℃ and retention time of 3 h were carried out for a series of vacuum pressure of 100000 Pa, 6000 Pa, 3000 Pa, 1500 Pa, 1000 Pa and 500 Pa.

Fig. 2-17 shows the relations between the concentration of fluorides and cyanides in the residues and the vacuum pressure in the experimental system.

As shown in Fig. 2-17, the fluoride content in the residue after treatment significantly drops to 58.8 mg/L at 100 kPa from the high level of 2342 mg/L in the raw SCCB. Furthermore, the fluorides content in the treated SCCB is being quickly decreased to 25.3 mg/L and 17.8 mg/L when the vacuum pressure drops to 6000 Pa and 3000 Pa respectively. The further drop of vacuum pressure lead to the deep removal of fluorides and the soluble fluoride content in the treated sample is eventually maintained at a level

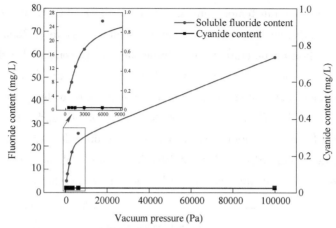

Fig. 2-17 Fluoride/cyanide removal in SCCB with different vacuum pressure at 1500 ℃, retention time 3 h

of 5–7 mg/L. The cyanide decomposes at the high temperatures and the cyanide content is kept below the lowest detection concentration (≤0.025 mg/L).

The conclusion comes out that the lower vacuum pressure in the system enhanced significantly the fluoride removal. After the system pressure is lower than 3000 Pa, it shows more powerful fluoride removal effect.

The heat treatment experiments at a lower temperature were carried out to obtain the detail of cyanide removal effects. Fig. 2-18 shows the cyanide content in the SCCB after treatment under the different temperatures.

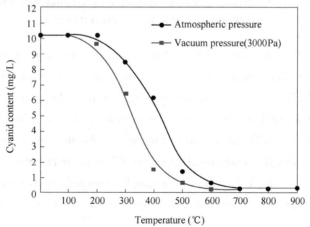

Fig. 2-18 Removal of cyanide in SCCB at 3000 Pa and atmospheric pressure

It can be seen in Fig. 2-18 that a large amount of cyanide was destroyed between 200–

500 ℃ at 100 kPa. As the temperature is raised to above 700 ℃, the cyanide in the SCCB is almost completely decomposed. A lower vacuum pressure of 3000 Pa makes the cyanide decomposition temperature reduced by about 100 ℃ compared with at 100 kPa. The cyanides can be completely decomposed when the temperature is higher than 600 ℃.

The heat treatment experiments were conducted at a series of temperatures 1100 ℃, 1200 ℃, 1300 ℃, 1400 ℃, 1500 ℃, 1600 ℃ and 1700 ℃ under the fixed vacuum pressure of 3000 Pa and retention time of 3 h to study the effect of temperature on the fluorides removal from SCCB. Fig. 2-19 shows the relationships between the concentration of fluorides in the residues and treatment temperature.

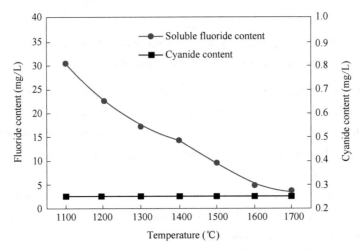

Fig. 2-19 Fluoride/cyanide removal in SCCB with different temperature at 3000 Pa, retention time 3 h

It can be seen from Fig. 2-19 that the fluoride content in SCCB decreases remarkably with the increasing temperature, which means that the better fluoride removal can be achieved at the higher temperatures. The fluoride content in the treated SCCB at 1100 ℃ is reduced to 30.2 mg/L compared with the fluoride content of 2342 mg/L in the raw SCCB indicating a significant defluoridation effect. As the temperature increases, the fluoride salt in the SCCB is continuously removed. When the temperature reached 1700 ℃, the concentration of soluble fluoride in the sample eventually decreased to 3.5 mg/L. In addition the cyanides is not yet detected in the treated SCCB (≤0.025 mg/L) at high temperatures.

It can be concluded that the increasing treating temperatures enhance the defluoridation, especially favoring the removal of the trace fluorides in the residual. An ideal defluoridation can be obtained at 3000 Pa at around 1700 ℃. Compared to the vacuum pressure functions the higher temperatures give a much better removal effect and the lower fluorides content in SCCB can be achieved. Finally, no matter which methods can be applied, i. e. increasing the vaccum or elevating the temperatures, such harmful substances in SCCB as fluorides and cyanide can be separated from carbon materials.

The weight loss of SCCB due to the volatilization of fluorides can be measured during the temperature-vacuum joint controlling process. Therefore the weight loss of the SCCB under atmospheric and vacuum pressures was compared to further indicate intuitively the discrepancy in the defluoridation effect under different vacuum pressures.

It can be seen from Fig. 2-20 that the weight loss of SCCB can be obtained under both atmopheric and vacuum pressures with the temperature increase. The weight loss rate of the SCCB is significantly increased between 1000 ℃ and 1200 ℃ at atmosphreic pressure, which is consistent with the theoretical volatile point of fluorides as shown in Table 2-5. However, a slow increase of the weight loss rate was exihibited as the temperature continues to be elevated at the temperature interval of 1200-1700 ℃. This means that some of the residual fluoride salt in the SCCB is slowly removed. These fluoride salts penetrate more closely in the SCCB and take longer to remove during the phase change. Fig. 2-21 shows a schematic diagram of the removal behavior of the fluoride salt from SCCB.

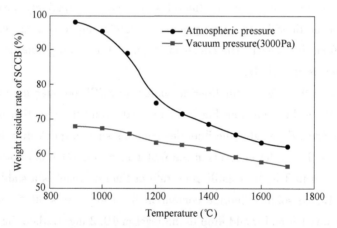

Fig. 2-20 Weight residue rate of SCCB at different temperatures at 3000 Pa and atmospheric pressure

(The retention time is 3 h)

In addition, the onset volatilization temperature is lowered as the vacuum pressure drops to a level of 3000 Pa. For instance, there is a 32.02% weigh loss at 900 ℃ under 3000 Pa compared to only 2.10% loss under the same temperature but atmosphreic

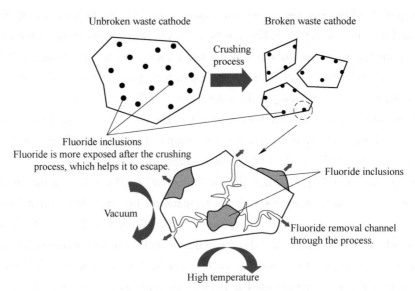

Fig. 2-21 Behaviour of volatilzation of fluorides through the porous of SCCB

pressure. It can be seen that the lower vacuum pressure lowers not only the volatilization point of fluorides, but also enhance the diffusion kinetics of fluorides gases and further make the fluorides volatilize more quickly.

The effects of retention time on fluorides removal is separately investigated. The volatilization of the fluorides is a slow process so maintaining a treating time is necessary for the better fluoride removal. The effects of retention time on the fluorides removal were also investigated in this study.

Fig. 2-22 shows the effect of the fluoride content in SCCB on retention time at 1500 ℃ and 3000 Pa. It can be seen from Fig. 2-22 that extending the retention time favors the removal of fluorides. Especially, the fluorides contents show a quick drop from the initial level of 2342 mg/L to a lower levels at the first 1 h. As the retention time is continually extended from 1 h to 3 h, the volatilization rate of fluoride reaches a stable state.

Quantitatively, the soluble fluoride content in the sample was 30.5 mg/L when the retention time was 1 h and would drop to the level of 10.2 mg/L when the retention time extends to 2.5 h and then kept almost unchanged for further extending time at 1500 ℃. The soluble fluoride content, at the level of 9.6–10.5 mg/L, was no longer decreased when the retention time was longer than 2 h.

Similar to the temperature and vacuum pressure the retention time plays also functions on fluoride removal. Longer retention time enhances the removal effects. It was considered from the experiment results that the ideal retention time is between 1.5 h and

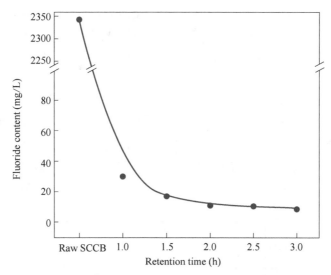

Fig. 2-22 The removal effect of fluoride in SCCB with extended retention time, at 3000 Pa, 1500 ℃

2.5 h for a SCCB with a particle size of 3–5 mm. Too long retention time will become not effective to improve the fluoride removal.

The effect of different test parameters on the detoxification of SCCB has been studied. A dimensionless analysis method was introduced into this study to further explore the correlation between the various factors (such as temperature, vacuum pressrue and retention time). The experimental data was mathematically fitted after being dimensionless, and the results are shown in the Fig. 2-23.

The details of the variables in the Fig. 2-23 are listed as follows: c is soluble fluoride content in the SCCB, mg/L; T is processing temperature, K; p is vacuum pressure, Pa; t is retention time, h; c_0 is 100 mg/L (the lower limit of hazardous waste); T_0 is 273 K; p_0 is 1×10^5 Pa; t_0 is 1 h.

In Fig. 2-23(a) the vacuum pressure and the removal effect of the fluoride are related in the form of a power function. It means that the fluoride removal effect in the SCCB becomes more significant as the treatment vacuum pressure drops. The fitting formula (Eq. (2-9)) is listed as follows:

$$\frac{c}{c_0} = 0.59 \left(\frac{p}{p_0}\right)^{0.38} \tag{2-9}$$

In Fig. 2-23(b) the removal effect of the fluorides is related with the temperatures in the form of a power function. It means that the fluoride removal effect in the SCCB becomes more significant as the treatment temperature increases. The fitting formula

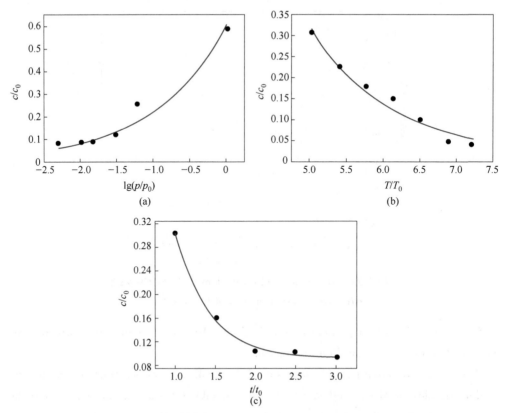

Fig. 2-23 The effect of different experimental conditions on fluoride removal
(a) Effect of p/p_0 on c/c_0; (b) Effect of T/T_0 on c/c_0; (c) Effect of t/t_0 on c/c_0

(Eq. (2-10)) is listed as follows:

$$\frac{c}{c_0} = 727.6\left(\frac{T}{T_0}\right)^{-4.8} \quad (2\text{-}10)$$

In Fig. 2-23(c) the removal effect of the fluorides is related with the retention time in the form of a natural logarithmic function. It means that the removal effect of fluorides in the SCCB first increases and then tends to be gentle with the prolongation of the retention time. The fitting formula (Eq. (2-11)) is listed as follows:

$$\frac{c}{c_0} = 2.2e^{-\frac{1}{0.43}\left(\frac{t}{t_0}\right)} + 0.09 \quad (2\text{-}11)$$

Based on the analysis of all kinds of influencing factors and abandoning the subordinate factors a simple fitted formula of the equivalent conversion can be presented. A correction factor $k = 35550$ was introduced into the data processing considering all the factors together. The final result of the mathematical function (Eq.

(2-12)) is listed as follows:

$$\frac{c}{c_0} = 35550 \left(\frac{T}{T_0}\right)^{-4.8} \left(\frac{p}{p_0}\right)^{0.38} \left[2.2 e^{-\frac{1}{0.43}\left(\frac{t}{t_0}\right)} + 0.09 \right] \quad (2\text{-}12)$$

The calculation error obtained by the above formula compared with the experiment results is less than 3.6 mg/L for SCCB with a particle size of 3-5 mm under the conditions of the temperature of 900-1700 ℃, the vacuum pressure of 500-100000 Pa and the retention time of 1-3 h. which cover almost all the changable ranges.

During the temperature-vacuum joint controlling treatment process all of the factors such as temprature, vacuum pressure and retention time play functions on the removal of fluorides and cyanides. The soluble fluoride content in SCCB can be reduced to a minimum of 3.5 mg/L and the cyanide be completely decomposed.

Fig. 2-24 shows the fixed carbon content of the SCCB treated under different temperatures of 1400 ℃, 1500 ℃, 1600 ℃ and 1700 ℃. The vacuum pressures were chosen to be 100000 Pa and 3000 Pa, and the retention time was 3 h for all cases.

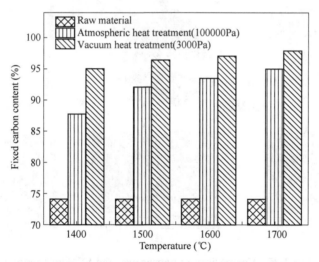

Fig. 2-24 Fixed carbon content of SCCB under different treatment conditions
(The retention time is 3 h)

The fixed carbon content in the SCCB treated increases with higher temperature and lower vacuum pressure. For instance, the fixed carbon content was 95.07% (wt) in the sample after atmospheric pressure treatment at 1700 ℃, while the fixed carbon content has reached 95.04% (wt) at the temperature 1400 ℃ under lower vacuum pressure of 3000 Pa. The lower vacuum pressure can reduce the treating temperature by about 300 ℃ for achieving the same level of fixed carbon content. The SCCB treated at 3000 Pa, 1700 ℃ had a maximum fixed carbon content of 97.89% (wt). Therefore, a joint temperature-vacuum controlling process can obtain a high quality carbon materials.

It should be noted that the products with a much higher carbon content can only be highly reutilized. For example, in the steel industry, the carbonaceous material with a fixed carbon content of greater than 95.46% can be considered as a coal-based recarburizer. The treating conditions of higher temperature than 1700 ℃ and lower vacuum pressure than 3000 Pa can meet such requirement needs.

The appearance and SEM micrograph observation of the raw SCCB, residue by treatment at 1700 ℃ and 3000 Pa for 2 h and fluorides collected in the condensation zone are compared in Fig. 2-25. It can be seen from Fig. 2-25(a) that the surface of the raw SCCB has been covered with an off-white electrolyte (fluorides). The electrolyte is tightly embedded in the voids and surface of the SCCB, and a large amount of dispersed fluorides are distributed in the pores in the form of irregular rod or cube shape as shown in Fig. 2-25(d).

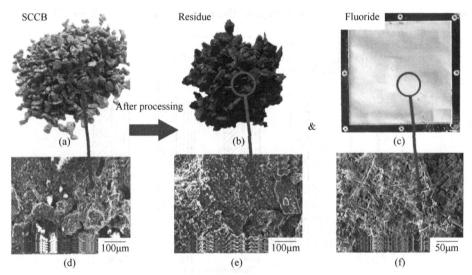

Fig. 2-25 Comparison in appearance and SEM micrograph
(a)(d) Raw SCCB; (b)(e) Residue by being treated at 1700 ℃, 3000 Pa, 2 h;
(c)(f) Fluoride collected in the condensation zone

After high temperature treatment the off-white fluorides covered on the surface of the SCCB disappears and the dark black carbonaceous material appears as shown in Fig. 2-25(b). Meanwhile, The SEM micrograph of the treated SCCB shown in Fig. 2-25(e) has a more regular morphology state and the carbon substrate becomes a fine spherical carbonaceous structure. Those fluorides that were previously distributed in the pores have also disappeared in Fig. 2-25(e). The fluorides volatilized from the SCCB and collected in the condensation zone are shown in Fig. 2-25 (c). It is shown that the collected fluorides appear as the pale yellow crystal.

The XRD pattern of the raw SCCB, residue by being treated at 1700 ℃ and 3000 Pa for 2 h, and the fluorides collected in the condensation zone are compared in Fig. 2-26.

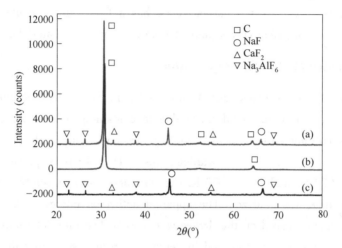

Fig. 2-26 Comparison in XRD pattern
(a) Raw SCCB; (b) Residue by being treated at 1700 ℃, 3000 Pa, 2 h;
(c) Fluoride collected in the condensation zone

Fig. 2-26(a) shows that there is a large amount of fluorides in the SCCB except the carbonaceous material, which exist in the form of sodium fluoride (NaF), calcium fluoride (CaF_2) and cryolite (Na_3AlF_6). The heat-treated SCCB residue is mainly carbonaceous material as shown in Fig. 2-26 (b) and the previous fluoride has been volatilized into the condensation zone after being treated mainly in the form of NaF and Na_3AlF_6.

The SCCB is considered to be a hazardous material in the aluminum reduction industry due to containing a large amount of soluble fluoride salt and toxic cyanide. The serious environment pollution and ecological destruction will be caused if the SCCB are only openpiled or buried in the smelters without any treatment.

A joint temperature-vacuum controlling process for treating SCCB is proposed. The thermodynamic analysis, experimental investigation and characteristics tests on the products material after treatment were carried out. The following conclusions have been drawn:

(1) Elevating the treatment temperature, decreasing the vacuum pressure and prolonging the retention time will improve the detoxification of the SCCB.

(2) The optimum reaction conditions can be set as the treatment temperature of 1700 ℃, the vacuum pressure of 3000 Pa and the retention time of 2 h.

(3) Under the optimal reaction conditions, the soluble fluoride content in the carbon material decreased from 2432 mg/L to 3.5 mg/L, cyanide was completely decomposed (≤0.025 mg/L). The treated carbon block has a fixed carbon content of 97.89% (wt). Fluoride can be effectively separated from carbonaceous materials.

2.5.2　Kinetics of fluorine separation

The spent cathode carbon block (SCCB) produced from the overhaul aluminum reduction cell is defined as hazardous solid waste due to containing a large amount of soluble fluoride and small amount of cyanide. Coupling vacuum-heating method is considered an effective way to deal with such hazardous waste. The SCCB is processed under the conditions of 3000 Pa and 1673-1973 K. The correlation between soluble fluoride content and treatment time was analyzed. The apparent activation energy of the detoxification reaction is calculated based on the *Arrhenius* equation. The rate determining step of the detoxification process is also analyzed in detail. The results show that the detoxification process is a first order reaction. Apparent activation energy $k = 15.08$ kJ/mol, and the rate determining step of the detoxification reaction is controlled by gas diffusion.

Almost all primary aluminum worldwide is produced by the Hall-Héroult process. The alumina and fluoride salt is respectively added into cryolite melt as raw material and additive. The average life of a reduction cell is usually 5-7 a. The spent lining materials (SPL) generated during the overhaul of the reduction cell is defined as hazardous solid waste due to containing the large amount of soluble fluoride and small amount of cyanide. Soluble fluoride and cyanide are easy to be leached into soil and groundwater when those hazardous wastes stacked in the open air. Excessive fluoride can cause damage to human bones, while grain yields can be significantly reduced when irrigated with fluoride-containing wastewater.

Some explorations on the harmlessness and resource utilization of the hazardous waste of the aluminum industry have been carried out in recent years[47-49]. Those treatment processes can be divided into hydrometallurgical and pyrometallurgical methods. Most of the hydrometallurgical methods are carried out in a liquid environment. Soluble fluoride is leached as much as possible to separate and recover the fluoride from the carbonaceous material. For example, Alcan uses LCL&L process to deal with SPL. A low-concentration NaOH solution was used to extract the fluoride and cyanide, and the fluoride ions in the filtrate were then precipitated by slaked lime. The cyanide contained in the filtrate was destroyed by hydrolysis at 180 ℃. The problem of SPL resource utilization can be partially solved by this method, but there are also obstacles such as long process flow and large investment in the actual operation process.

The pyrometallurgical method cannot recycle carbon materials with high value, as they will be burned. During the combustion process, harmful gases containing fluorine will be generated. The cyanide is decomposed as well. Reynolds uses a high-temperature method to deal with SPL[50]. Natural gas is used as a heat source, limestone and anticaking agents are added to the kiln. Cyanide is decomposed at high temperatures, and soluble fluoride salts are converted into insoluble calcium fluoride. The toxicity of SPL can be addressed by this method. However, a lot of valuable resources in the SPL are wasted, and a larger amount of useless residue is generated.

According to the literature, the hydrometallurgical method is a cost intensive and time consuming process. And during the process, it will be accompanied by the precipitation of HF and HCN, causing severe corrosion of the equipment. It is difficult to completely separate the products after treatment, and the pollution problem is serious. The traditional pyrometallurgical method cannot recycle the carbonaceous materials in the SCCB, and at the same time, a large amount of unwanted residue is generated.

Vacuum metallurgy technology has been continuously developed in recent years. Its applications include, but are not limited to, metal smelting. It can also be applied to the treatment of hazardous solid waste. Vacuum technology has the advantages of short processing time, low required temperature and high product purity compared to traditional processing methods. Some scholars have made some useful explorations on the vacuum treatment of the SCCB[51]. Knowledge about the kinetics research on the vacuum detoxification process of the SCCB is lacking. The variation of fluoride content in spent cathode carbon block with vacuum heat treatment time is investigated. The apparent activation energy of the detoxification reaction is calculated based on the Arrhenius equation. The rate determining step of the detoxification process is also analyzed in detail.

The SCCB used in this experiment came from a cell with a life of 2643 d in a smelter in western China. The electrolyte and alumina precipitates on the surface of the SCCB were first cleaned, and the bonded refractory materials were also sorted out. The sorted SCCB is crushed to 100-200 mm and then sealed in a sample bag to keep it dry. Before the experiment, these SCCB were crushed to 3-5 mm and mixed as much as possible to ensure uniformity. The fixed carbon content of SCCB is 74.08%, NaF content is 9.72%, Na_3AlF_6 content is 4.86%, and CaF_2 content is 2.43%.

The vacuum detoxification equipment used in the experiment is shown in Fig. 2-27. High-purity molybdenum rods are used as heating elements, and alumina ceramic fibers are used as refractory materials. The rated temperature range of the equipment is from room temperature to 1800 ℃. The vacuum in the furnace can be kept constant through the vacuum control panel in the lower right corner.

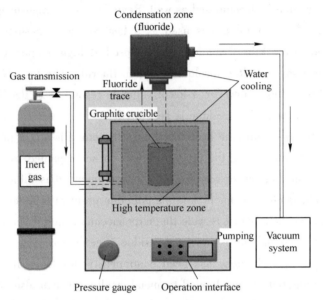

Fig. 2-27   Schematic diagram of vacuum detoxification equipment

For each test, the crushed SCCB (3 - 5 mm) was put into a graphite crucible and placed in the high temperature zone, and the detoxification kinetics research experiment was performed under a vacuum condition of 3000 Pa. The heating rate was 8 K/min at the temperatures lower than 1473 K and 5 K/min when temperature was higher than 1473 K. When the SCCB is heated to a preset temperature with the set curve, the heating is stopped immediately after maintaining a certain retention time. At this time, argon gas was introduced into the furnace to accelerate cooling, and the cooling water system of the equipment was also turned on to ensure that the retention time of each test was strictly in accordance with the set value. The detoxification temperature is set as 1673 K, 1773 K, 1873 K and 1973 K. The retention time is set to 1 h, 1.5 h, 2 h, 2.5 h and 3 h.

The soluble fluoride in the treated samples was detected by fluoride ion-selective electrode method in GB/T 15555.11—1995. The decomposition of cyanide in SCCB has been described in detail in our previous work, so it will not be discussed. Further, the dynamic process of SCCB vacuum heat treatment is studied in detail.

Fluoride can be separated from the SCCB by means of vacuum heating. In our previous research, the fluoride recovered in the condensation zone was found to consist of sodium fluoride, calcium fluoride and cryolite. Interestingly, this is exactly the same as the composition of fluoride in SCCB. There is no chemical reaction between the several fluorides, and no new compounds are produced. This may indicate that the removal of fluoride under vacuum is a phase transition process. Further, the reaction formula of the detoxification process of SCCB is shown in Eq. (2-13)-Eq. (2-14).

$$NaF(s) + CaF_2(s) \xrightarrow{\text{Vacuum heating}} NaF(g) + CaF_2(g) \xrightarrow{\text{Condensation}} NaF(s) + CaF_2(s) \quad (2\text{-}13)$$

$$Na_3AlF_6(s) \xrightarrow{\text{Vacuum heating}} NaF(g) + AlF_3(g) \xrightarrow{\text{Condensation}} Na_3AlF_6(s) \quad (2\text{-}14)$$

Normally, cryolite is difficult to be decomposed. However, cryolite can be decomposed into sodium fluoride and aluminum fluoride under vacuum heating. Sodium fluoride and aluminum fluoride will be converted back to cryolite in the subsequent condensation process. The Gibbs free energy of the above reaction is calculated. As shown in Fig. 2-28, the thermodynamic calculation results are consistent with the experimental results, and cryolite was also found in the condensate.

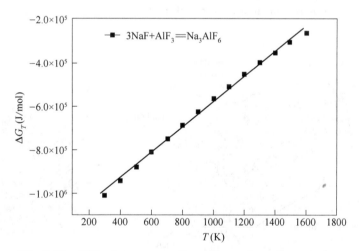

Fig. 2-28 Gibbs free energy of cryolite condensation reaction

The detoxification process of SCCB can be composed of the following steps: (1) Fluoride migrated from the inside of the SCCB to the liquid phase boundary layer. (2) Volatilization at the liquid/gas phase interface. (3) The volatile fluoride diffuses into the gas phase through the gas boundary layer. (4) The volatile fluorides migrate to the condensing wall in the gas phase. (5) Condensation of the volatile fluoride.

Step (3) may become the rate determining step of the whole process when the gas phase pressure above the SCCB is high. However, vacuum conditions were applied in this experiment, and the element diffusion in the gas phase boundary layer should not be a rate determining step. Mass transfer in the gas phase is much faster than in the liquid phase, and steps (4) and (5) do not become the rate determining step. The gaps and holes in the SCCB are complex, so step (1) may become the rate determining step. Further, liquid fluoride requires some energy to overcome surface tension during the phase transition. Therefore, the volatilization reaction of fluoride at the liquid/gas interface may also be the rate determining step.

During the years of service of the cathode carbon block, molten salt electrolyte continued to penetrate into it, and sodium fluoride, cryolite and calcium fluoride accumulated in the SCCB. The schematic diagram of SCCB detoxification process is shown in Fig. 2-29. These fluoride salts deposited in the microspores of SCCB undergo a phase transition reaction during the vacuum heat treatment. Phase-changed fluorides are transferred to the surface along the path where they originally penetrated into the SCCB or other channels inside the SCCB that are connected to the outside as the detoxification process proceeds. However, these pores and channels are usually narrow and complex, which makes the removal of fluoride very difficult.

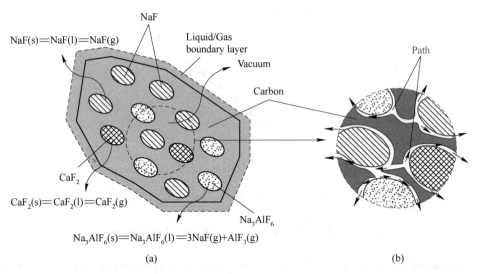

Fig. 2-29 Schematic diagram of fluoride removal process in the SCCB

(a) Spent cathode carbon block; (b) Fluoride removal path

A series of tests were used to verify the above-mentioned conjecture about the detoxification mechanism. The relationship between the residual ratio of soluble fluoride in SCCB and the retention time after treatment at different temperatures under vacuum conditions (3000 Pa) is shown in Fig. 2-30. The initial soluble fluoride concentration in the SCCB is 2342 mg/L.

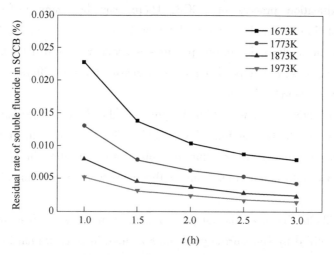

Fig. 2-30 Residual ratio of soluble fluoride in SCCB after different temperature and time treatment

As shown in Fig. 2-30, the residual ratio of soluble fluoride in SCCB treated with different temperatures showed a gradually decreasing trend with the extension of the retention time. In other words, the soluble fluoride remaining in the SCCB is gradually removed by volatilization. It is worth noting that the decreasing trend of residual fluoride concentration in SCCB became gentler when the retention time was extended from 2 h to 3 h. The cause of this phenomenon may be that the fluoride remaining in the SCCB at the later stage of the detoxification process is the most tightly bound, or those that penetrate into the deepest gap of the SCCB. For these stubborn soluble fluorides, a longer path and a stronger driving force are needed to make them transfer from the interior of the SCCB to the surface during the detoxification process.

Higher temperatures can provide stronger power for the removal of these fluorides. As shown in Fig. 2-30, the soluble fluoride concentration of the treated sample decreased with increasing treatment temperature when the retention time was the same. At a residence time of 1 h, the residual ratio of soluble fluoride in the sample treated at 1673 K was 0.023% (the concentration of soluble fluoride was 53.3 mg/L). However, with the same residence time, the residual ratio of soluble fluoride in the sample treated at 1973 K was 0.005% (the concentration of soluble fluoride was 12.3 mg/L). Under 3000 Pa, both elevated temperature and extended residence time can promote the detoxification of SCCB. Further, the effect of increasing the temperature is more significant.

At present, no scholars have carried out detailed studies on the dynamics of the vacuum detoxification process of SCCB. There may be many restrictions on the volatilization of fluoride in the vacuum detoxification process of SCCB. The theoretical calculations of thermodynamics in our previous research work did not take into account the above restrictions, which is also the main reason for the differences between the experimental and theoretical values.

In order to obtain a more efficient process, the kinetic mechanism of SCCB detoxification needs to be carefully investigated. The reaction rate constant, reaction order, and activation energy of the fluoride volatilization reaction in SCCB need to be obtained, and the rate determining step of the detoxification reaction also need to be found out.

In 1850, Wilhelm first used the integral method to measure the reaction order, which was later generalized by Harcourt and Esson. For fluoride volatilization in the SCCB, the reaction rate equation for $[F]_{carbon} = F(g)$ is shown in Eq. (2-15).

$$-\frac{dc_F}{dt} = kc_F^n \quad (2\text{-}15)$$

Where d represents the differential symbol; t is the retention time; c_F is the concentration of soluble fluoride in the SCCB after t time vacuum treatment; n is the reaction order; k is the reaction rate constant.

$$-\frac{dw_F}{dt} = kw_F^n \quad (2\text{-}16)$$

When the concentration of soluble fluoride in SCCB is expressed by mass fraction w_F, the reaction rate equation Eq. (2-15) can be written as Eq. (2-16).

As shown in Fig. 2-31, the experimental results were used to plot the reaction order ($n = 0, 1$ and 2) in Eq. (2-16). It is found that when $n = 1$ and $n = 2$, a good linear fit can be obtained.

In fact, many reactions have been found to obtain good correlations simultaneously when the reaction order is $n = 1$ and $n = 2$. Therefore, it is one-sided to explore the order of the reaction through the correlation of the reaction order image. In order to accurately determine the order of the SCCB detoxification reaction, a half-life method was introduced to determine the order of the reaction (Eq. (2-17)). Where X_F is the conversion rate of fluoride in SCCB at time t.

$$X_F = \frac{c_{F,0} - c_F}{c_{F,0}} \quad (2\text{-}17)$$

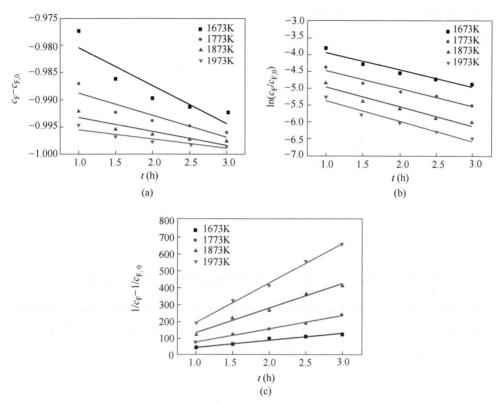

Fig. 2-31 Experimental data fitting under different reaction order
(a) $n=0$; (b) $n=1$; (c) $n=2$

Then the reaction rate equation can be expressed as Eq. (2-18).

$$k_F = \frac{1}{t}\ln\frac{1}{1-X_F} \quad (2\text{-}18)$$

When the soluble fluoride conversion rate in SCCB is 50%, that is, when $X_F = 0.5$, the half-life of the first-order reaction is $t_{1/2}$. As shown in Eq. (2-19).

$$t_{1/2} = \frac{\ln 2}{k} \quad (2\text{-}19)$$

Similarly, the half-life of the second-order reaction is $t'_{1/2}$ as shown in Eq. (2-20).

$$t'_{1/2} = \frac{1}{kc_{F,0}} \quad (2\text{-}20)$$

The calculation results of half-life at different temperatures are shown in Fig. 2-32. When the reaction order $n=1$, the half-life of the detoxification process of the SCCB at the temperature of 1673–1973 K is $t_{1/2} = 1.13$–1.33 h. However, the half-life of the detoxification process of the SCCB at the temperature of 1673–1973 K is $t'_{1/2} = 0.004$–

0.023 h when the reaction order $n = 2$.

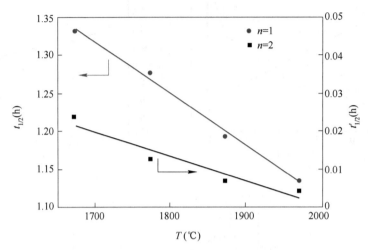

Fig. 2-32 The half-life of the detoxification process when $n = 1$ and $n = 2$

Combined with the experimental data, the half-life of the reaction order $n = 1$ obviously has higher reliability. It indicates that $[F]_{carbon} = F(g)$ reaction is a first-order reaction. The reaction rate equation is shown in Eq. (2-21).

$$\ln \frac{w_{F,t}}{w_{F,0}} = -kt \tag{2-21}$$

Therefore, the slope of the linear fitted line at the temperatures of 1673 K, 1773 K, 1873 K and 1973 K is the rate constant k of the reaction at the corresponding temperature. For 1673 K, the reaction rate constant $k = 1.44 \times 10^{-4}$ s^{-1}. For 1773 K, the reaction rate constant $k = 1.51 \times 10^{-4}$ s^{-1}. For 1873 K, the reaction rate constant $k = 1.61 \times 10^{-4}$ s^{-1}. For 1973 K, the reaction rate constant $k = 1.70 \times 10^{-4}$ s^{-1}.

The reaction rate constant k and the experimental temperature T obtained from the experiments usually conform to the *Arrhenius* formula Eq. (2-22). Using the *Arrhenius* formula to calculate the apparent activation energy to find out the rate determining step is one of the common methods for studying reaction kinetics.

$$k = Ae^{-\frac{E_a}{RT}} \tag{2-22}$$

where, A is the former factor, and its unit is the same as the rate constant; E_a is the apparent activation energy of the reaction, kJ/mol; T is the reaction temperature, K; R is the molar gas constant, 8.314 J/(mol · K).

Eq. (2-22) can be transformed to Eq. (2-23):

$$\ln k = -\frac{E_a}{RT} + \ln A \tag{2-23}$$

In Eq. (2-23), $\ln k$ and $\dfrac{1}{T}$ have a linear relationship. Plot $\ln k$ and temperature $\dfrac{1}{T}$ at temperatures of 1673 K, 1773 K, 1873 K, and 1973 K, as shown in Fig. 2-33. By calculating the slope of the linear fitted line, the activation energy of the $[F]_{carbon} = F(g)$ reaction is $E_a = 15.08$ kJ/mol.

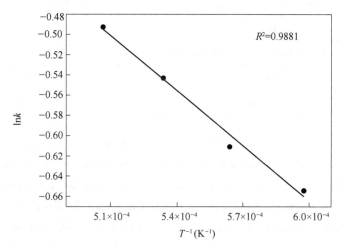

Fig. 2-33 Relationship between detoxification rate constant k and temperature T

The rate determining step of fluoride volatilization in SCCB has not been studied by scholars before. Savov uses activation energy as a criterion to determine the volatility-limiting link in the melt, which can be introduced into this study[52]. If the apparent activation energy is close to the value of the enthalpy of evaporation of the pure substance, the chemical reaction at the interface between the liquid and gas phases is the rate determining step. However, if the apparent activation energy is much less than the value of the enthalpy of evaporation of the pure substance, the chemical reaction at the interface between the liquid and the gas phase is not the main rate determining step.

The enthalpy of evaporation of pure NaF at its boiling point (1986 K) is 135.69 kJ/mol. The apparent activation energy of the $[F]_{carbon} = F(g)$ reaction in this experiment is 15.08 kJ/mol. Apparent activation energy is much smaller than the enthalpy of volatilization of pure substances, so the volatilization reaction at the gas/liquid phase interface is not the rate determining step of detoxification process.

Table 2-8 shows the corresponding relationship between the reaction activation energy and the rate determining step. Generally, the reaction process is under kinetic control when $E \approx 40-300$ kJ/mol, and the reaction process is under diffusion control when $E \approx 8-20$ kJ/mol. In this experiment, the apparent activation energy of the detoxification process of the SCCB is in the range of gas diffusion, so the migration of fluoride from the inside of the SCCB to the liquid boundary layer may be the rate determining step.

Table 2-8 The relationship between reaction activation energy and the rate determining step

The rate determining step	Reaction Activation Energy(kJ/mol)
Gas diffusion	8-16
Gas diffusion & Interfacial chemistry reaction	29-42
Interfacial chemistry reaction	60-67
Solid diffusion	>90

According to the dynamic process analysis results, this indicates that the rate determining step of the fluoride removal process in SCCB is the gas diffusion process.

The detoxification efficiency can be effectively improved by vacuuming the reaction system. The vacuum environment outside the SCCB allows the fluoride transferred from the inside to the surface to be continuously removed and collected by condensation. Therefore, increasing the reaction temperature can enhance the driving force of fluoride movement, and vacuuming the system is conducive to the removal of fluoride on the surface of SCCB. Both methods mentioned above are beneficial to improve the rate of SCCB detoxification process.

SCCB is a hazardous solid waste generated during the production of primary aluminum. It contains a large amount of soluble fluoride and high-value carbonaceous materials, which has extremely high potential recycling value. The variation of fluoride content in spent cathode carbon block with vacuum heat treatment time is investigated. The apparent activation energy of the detoxification reaction is calculated based on the Arrhenius equation. The rate determining step of the detoxification process is also analyzed in detail. The following conclusions have been drawn:

The fluoride volatilization reaction in the vacuum heat treatment process of the SCCB $[F]_{carbon} = F(g)$ is a first-order reaction, and the apparent activation energy of the reaction is 15.08 kJ/mol. The volatilization process rate is mainly controlled by the fluoride migration from the interior of the SCCB to the liquid phase boundary layer. The reaction rate constants from 1673 K to 1973 K are 1.44×10^{-4} s^{-1}, 1.51×10^{-4} s^{-1}, 1.61×10^{-4} s^{-1} and 1.70×10^{-4} s^{-1}, respectively. The fluoride and carbonaceous materials in SCCB can be effectively separated by vacuum heat treatment, and the removal of fluoride can be promoted by increasing the temperature and extending the retentione time.

2.5.3 Thermal conductivity simulation of waste cathode carbon block

China is a major producer of primary aluminum. With the extension of the service life of aluminum electrolysis cells, the adsorption amount of fluorine in cathodes and refractory insulation materials is also increasing year by year. Generally, carbon bricks and refractory insulation bricks need to be replaced after 5-8 a of use, resulting in a large amount of solid waste. Statistics show that primary aluminum plants produce an average of 20-30 kg of waste cathode per 1 t of aluminum production[54]. Carbon accounts for 30% to 70% of the waste cathode, with a graphitization degree of up to 80% to 90%. The remaining substances mainly include valuable components such as cryolite, sodium fluoride, lithium fluoride, etc.[55]. Waste cathode carbon blocks from aluminum electrolysis cells are classified as hazardous solid waste due to their presence of soluble fluorine and trace cyanide. They can react violently with water at room temperature and pressure, and emit harmful gases. Improper disposal can also cause serious ecological damage. The national policy stipulates that the storage time of hazardous solid waste shall not exceed one year. Due to environmental pressure and the demand for sustainable development of the aluminum industry, the harmless and resourceful treatment technology of waste cathode carbon blocks has attracted attention both domestically and internationally.

At present, the treatment technology for hazardous waste from aluminum electrolysis at home and abroad mainly remains in the stage of on-site anti-seepage storage, landfill, or low-cost harmless and low-value treatment[56]. The treatment for harmless or low value treatment can be basically summarized as fire method and wet method. In terms of pyrometallurgical process, the waste cathode Osmet treatment technology jointly developed by Aluminum Corporation of the United States and Osmet Company of Australia[57] recovers fluorine in the form of aluminum fluoride, directly burns the carbon, and incinerates the slag for road construction; France's Bischner Aluminum has developed a pyrolysis process by adding mineral additives to the spent potlining[58]; Puji Aluminum Canada[59] mixes the crushed and ground waste cathode carbon block with calcium sulfate, and then solidifies the fluoride through high-temperature calcination; In 1992, Cormac Aluminum Corporation of Australia reported the Comalco SPL treatment process, which had already passed the experimental stage. In the material preparation stage, the waste electrolytic cell lining material was crushed into particles with a particle size of less than 1 mm, and fluoride was recovered after calcination; Zhengzhou Research Institute of Aluminum Corporation of China[60] used waste cathode carbon blocks as raw materials and fly ash rich in SiO_2 and Al_2O_3 as reaction dispersants. After roasting, sulfuric acid and lime were used to decompose at room temperature, ultimately achieving harmless treatment of waste cathode carbon blocks. All of the above processes can achieve the harmless treatment of waste cathodes, but they do not attach importance to the recycling and utilization of valuable substances. Realizing industrialization economically still faces many challenges. Wet treatment mainly includes flotation method[61], high-temperature hydrolysis method[62], and sulfuric acid treatment method. The main process technology that has initially achieved industrialization is a comprehensive wet treatment of spent potlining(including waste cathode carbon blocks) developed by Rio Tinto Alcan abroad. This method was put into operation in April 2008 and reached its maximum production in 2014. However, it has been facing problems such as low resource utilization and easy secondary pollution. In 2013, Beijing Mining and Metallurgy Research Institute and China Power Investment Ningxia Energy Aluminum jointly developed a wet treatment technology that comprehensively utilizes waste cathode carbon blocks and waste refractory materials. Valuable carbon can be effectively recovered, but the process is long and the cost of wastewater treatment is high. Generally speaking, wet treatment has problems such as long process flow and secondary pollution caused by harmful substances entering the solution. However, the pyrometallurgical method has not achieved effective recovery and utilization of valuable components such as carbon and fluoride salts, and its industrialization path is still being explored. In addition, the treatment of waste cathode carbon blocks is also used in other industries such as steelmaking and cement manufacturing[63-64]. However, due to the presence of harmful substances such as sodium and fluorine in waste cathode carbon blocks, as well as strict environmental monitoring measures in various countries, its promotion and application are limited.

Based on the current large inventory of waste cathode carbon blocks in China, low consumption of existing technologies, and low degree of resource utilization, this article proposes an electric heating high-temperature heat treatment technology for waste cathode carbon blocks in aluminum electrolysis cells. By utilizing the characteristics of high conductivity, high graphitization degree, and high volatility of high-temperature fluoride salts in waste cathode carbon blocks, under the condition of electric heating in a high-temperature furnace, the goal of efficient separation of carbon and fluoride salts, and recovery of valuable components is achieved, And a set of experimental scale electric heating high-temperature heat treatment furnace has been established. After determining the temperature required for effective volatilization of particulate fluoride salts based on experiments, numerical simulation is used to study the temperature evolution law of high-temperature resistance furnaces, optimize and adjust process control parameters, provide reasonable power supply curves, and provide reference for industrial heating furnace process control. It should be noted that for the simulation calculation of electric heating high-temperature resistance furnace, this article only considers the heating and insulation stages, and the cooling process and fluoride salt cooling recovery process will be further studied in the future.

Based on heat transfer calculations and referring to furnace type parameters such as Acheson graphitization furnace and silicon carbide furnace, a high-temperature resistance furnace as shown in Fig. 2-34 was designed, mainly consisting of a furnace, furnace core, electrodes, and insulation materials, with each part being square. The insulation material includes the upper layer insulation material and the surrounding fire-resistant insulation material, among which the upper layer insulation material serves as the furnace cover, including three parts: carbon plate, graphite felt, and rock wool; The surrounding fire-resistant insulation material consists of five parts: carbon bricks, fillers, high alumina bricks, insulation bricks, and rock wool. One power supply electrode on each side, running through the refractory insulation layer, with the electrode center 790 mm from the bottom of the furnace type. The furnace core is located in the center of the furnace. The specific size parameters of each part are shown in Table 2-9, and Fig. 2-35 shows the detailed insulation structure of each layer in the furnace.

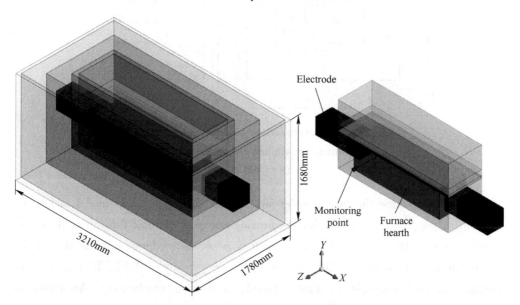

Fig. 2-34 Schematic of the high temperature resistance furnace and its interior detail

Table 2-9 Structure size of the high temperature resistance furnace   (mm)

Parameters	Numbers
Length	3210
Height	1680
Width	1780
Thickness of carbon brick	230
Thickness of insulating filler	65
Thickness of high alumina brick	230
Thickness of insulating brick	230
Thickness of rock wool	50
Dimensions of electrode	300×300×805
Dimensions of furnace hearth	500×500×1600
Dimensions of furnace core	50×50×1600

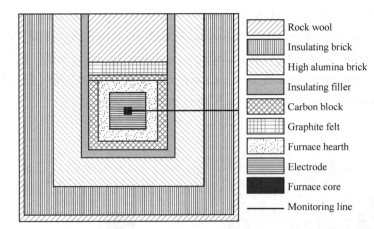

Fig. 2-35 The cross-section of the high temperature resistance furnace and the monitoring line
($X = 1.605$ m)

In order to describe the heat transfer situation in the furnace, this model makes the following assumptions: Assuming that the furnace charge is isotropic throughout the entire heating process. The material in the furnace is a porous medium, and the size of the waste cathode carbon block is 30–70 mm, with a set porosity of 0.52; The heat transfer between stacked materials is mainly based on thermal conductivity. According to Russell's empirical formula[65], the effective thermal conductivity coefficient is calculated to be 6 W/(m · K); assuming that material density, thermal conductivity, specific heat capacity, and resistivity do not vary with temperature; Convection and radiation heat transfer occur on the outer wall, and the comprehensive heat transfer coefficient is taken as 14.31 W/(m² · K).

The control equations of the mathematical model mainly include heat transfer and electric field control equations, which are expressed as follows: The heat transfer in the calculation area is mainly transmitted through thermal conduction, so the heat transfer control equation is the Fourier thermal conduction differential equation:

$$\rho c_p \frac{\partial T}{\partial t} = \frac{\partial}{\partial x}\left(\lambda \frac{\partial T}{\partial x}\right) + \frac{\partial}{\partial y}\left(\lambda \frac{\partial T}{\partial y}\right) + \frac{\partial}{\partial z}\left(\lambda \frac{\partial T}{\partial z}\right) + q_s \qquad (2\text{-}24)$$

where ρ is the bulk density, kg/m³; c_p is the specific heat capacity, J/(kg · K); λ is the effective thermal conductivity, W/(m · K); q_s is the intensity of the heat source (heat generation rate per unit volume), W/m³; T is the absolute temperature, K; t is time, s; As coordinates, x, y, z is the coordinate, the values of various physical parameters are shown in Table 2-10.

Table 2-10 Thermophysical properties used in this study

Parameters	Density (kg/m³)	Specific heat (J/(kg · K))	Conductivity (W/(m · K))	Conductivity (S/m)
Carbon block	1500	800	10	1×10^{-10}
Insulating filler	450	1465	0.65	1×10^{-10}
High alumina brick	1500	1100	0.786	1×10^{-10}
Insulating brick	1000	1200	0.3	1×10^{-10}
Rock wool	135	900	0.06	1×10^{-10}
Graphite felt	1400	900	0.01	1×10^{-10}
SCCB	850	840	6	2.127×10^{3}
Electrode	1600	700	12.4	1.13×10^{-5}

The Joule heat generated by the current in the conductive zone is the main source of heat source intensity:

$$q_s = \tau |\nabla \varphi|^2 \qquad (2\text{-}25)$$

where τ is the resistivity, Ω · m; φ is the potential, V.

The electric potential control equation is as follows:

$$\frac{\partial}{\partial x}\left(\gamma \frac{\partial \varphi}{\partial x}\right) + \frac{\partial}{\partial y}\left(\gamma \frac{\partial \varphi}{\partial y}\right) + \frac{\partial}{\partial z}\left(\gamma \frac{\partial \varphi}{\partial z}\right) = 0 \qquad (2\text{-}26)$$

where γ is the conductivity of the material, S/m; φ is the potential, V.

Convective heat transfer occurs between the outer surface of the furnace sidewall and bottom and the workshop environment.

$$-\lambda \frac{\partial T}{\partial n} = h_s(T - T_e) \quad (2\text{-}27)$$

where λ is the thermal conductivity, W/(m·K); T is the wall temperature, K; T_e is the ambient temperature, K; h_s is the comprehensive convective heat transfer coefficient, W/(m²·K).

Comprehensive heat transfer includes two parts: convective heat transfer and radiation heat transfer:

$$h_s = h_c + h_r \quad (2\text{-}28)$$

where h_c is the convective heat transfer coefficient, W/(m²·K); h_r is the radiation heat transfer coefficient, W/(m²·K);

$$h_c = Nu(\lambda/L) \quad (2\text{-}29)$$

where λ is the thermal conductivity, W/(m·K); L is the qualitative dimension in the direction of heat transfer; Nu is the Nusselt number.

$$h_r = \varepsilon\sigma(t_1^4 - t_0^4)/(t_1 - t_0) \quad (2\text{-}30)$$

where ε is the blackness of the radiating object; σ is the Stephen Boltzmann constant, 5.67×10^{-8} W/(m²·K⁴); t_1 and t_0 are the wall temperature and ambient temperature, respectively, K.

The heat transfer of refractory insulation layer and materials in the furnace follows Fourier's law of thermal conductivity.

$$q = -\lambda_i \frac{dt}{dx} = \lambda_i \frac{\Delta t_i}{\delta} \quad (2\text{-}31)$$

where q is the heat passing through the wall surface of the insulation layer, W/m³; λ_i is the thermal conductivity of the i-th layer (5 layers of refractory insulation materials and 3 layers of upper insulation materials), W/(m·K); Δt_i is the temperature difference between the walls on both sides of the i-th layer, K; δ is the thickness of the insulation layer, m.

The boundary conditions of the electric field are set with the anode as the variable of the power supply potential, the cathode as the zero potential, and all other surfaces as insulation surfaces.

The volatilization temperature node of fluoride salts was determined using experimental methods. The waste cathode carbon blocks used in the experiment were from Qingtongxia Aluminum Industry Group, and their main components were detected through chemical analysis as shown in Table 2-11.

Table 2-11 The composition and the content of various elements in SCCB (%)

Elements	C	F	Na	Al	Ca	Fe	Others
Content	59.2	16.73	13.62	7.87	1.43	0.94	0.21

Use a jaw crusher to crush the waste cathode carbon blocks used in the experiment. Before each experiment, take 20 g samples with a certain particle size and load them into a crucible. Dry them in a 100 ℃ drying oven for 4 h and weigh them. Record data A. Firstly, nitrogen gas is introduced into the high-temperature energy-saving atmosphere furnace (with a maximum heat resistance temperature of 1800 ℃) for a period of time. Then, the dried sample is placed and the experimental temperature is set. After heating until the weight does not change, it is taken out and weighed, and data B is recorded. The ratio of the difference between the two weighings and the initial value A is the burning loss rate of the waste cathode carbon block.

The melting point of material carbon is relatively high, and during the heating process of waste cathode carbon blocks in aluminum electrolysis cells, fluoride salts are mainly evaporated and precipitated due to heating. According to the composition table of waste cathode carbon blocks used in the experiment, the carbon content before treatment accounted for 59.2%, and the total proportion of fluoride and other substances was 40.8%. As shown in Fig. 2-36, as the treatment temperature increases, the proportion of fluoride salt precipitation gradually increases, and the precipitation trend slows down when the temperature exceeds 1200 ℃. After treatment at 1600 ℃, the burning loss rate of the waste cathode carbon block is 36.73%, and about 90% of the fluoride salt

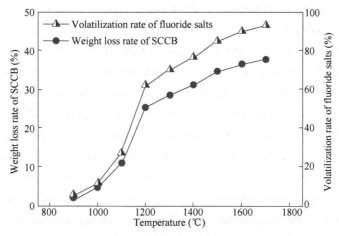

Fig. 2-36 Relationship between burning loss of SCCB and volatilization of fluoride salts at different temperatures

evaporates. 93.1% of fluoride salts evaporate after high-temperature treatment at 1700 ℃. Due to the high volatile point of CaF_2 and a small amount of non volatile iron containing substances in fluoride salts, 1700 ℃ is determined as the sufficient volatilization temperature for fluoride salts in waste cathode carbon blocks of aluminum electrolysis cells in this article.

The power-on period is set to 36 h. By simulating the temperature field in the high temperature furnace during the heating process, the temperature distribution in the furnace at different times and the heating curves at different positions of the furnace material and refractory insulation layer during the process were obtained. The voltage during the heating process(the first 24 h)was optimized and designed. The area with a temperature higher than 1700 ℃ in the furnace is defined as the effective volatilization area, and the proportion of the volume above 1700 ℃ to the total material volume is defined as the volatilization rate of fluoride salts. By integrating a specific physical quantity specified in the selected area of the model through the function function function in CFD-Post, the volume fraction of the physical quantity in different temperature ranges can be obtained. If the volatilization rate of fluoride salt is greater than 90%, it can be considered as complete volatilization.

The transfer of temperature from the center of the furnace to the edge takes a certain amount of time. By using a higher voltage to increase the maximum temperature at the center position, and utilizing the temperature difference, the time required for the overall heating of the furnace can be effectively reduced. If power transmission is maintained for a long time, there is a problem of energy waste, and excessive heating rate can affect thermal efficiency. The particularity of the furnace material dealt with in this article lies in its certain conductivity, which means that it will generate heat on its own when energized. In order to understand the changes in temperature inside the high-temperature furnace during the heating process, based on practical experience and theoretical calculations, a 12 V pressure supply for 24 h was selected, and then the heating insulation process was carried out under a 9 V pressure supply for 12 h.

Fig. 2-37 shows the temperature distribution cloud map corresponding to the cross-section of the furnace type($Z = 0.89$ m) under different power supply times. As can be seen in the figure, during the heating process, the furnace core and the furnace material together serve as the heating element, and the resistivity of the furnace material is much smaller than that of the furnace core. Therefore, the heating speed of the furnace core is fast and the thermal diffusion power is strong. During the heating process, the temperature gradually diffuses outward from the center, and the high-temperature

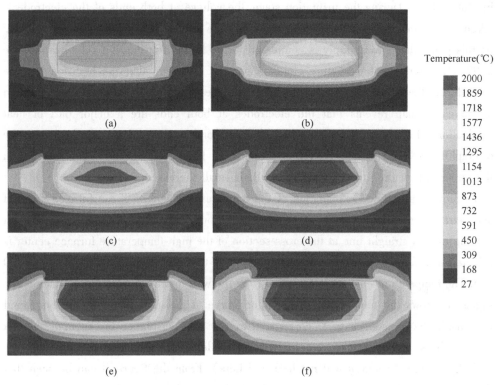

Fig. 2-37 Contours of temperature distribution in the cross section ($Z = 0.89$ m) of the furnace at different heating time

(a) 4 h; (b) 8 h; (c) 12 h; (d) 16 h; (e) 24 h; (f) 36 h

isothermal surface continuously diffuses outward. From the temperature distribution cloud map, it can be seen that after 12 h of power supply, the temperature of the furnace core and some furnace materials increases to around 2000 ℃. As the heating time prolongs, the high-temperature zone gradually expands. After heating for 18 h, the volatilization rate of fluoride salt is 83.1%; When the furnace core is heated for 24 h, the maximum temperature can reach 2250 ℃, and the volatilization rate of fluoride salt is calculated to be as high as 98%, meeting the requirement for complete volatilization of fluoride salt. As the furnace material heats up, the temperature of the insulation layer also continues to rise. Under higher voltage, the temperature of the furnace core rapidly increases, but certain insulation conditions are conducive to uniform heating of block materials. The waste cathode carbon blocks of aluminum electrolysis cells have a certain particle size, so they need to be kept at high temperature for a period of time after heating to ensure the heating rate and uniform temperature distribution in the furnace. To maintain the existing temperature level in the furnace while avoiding continuous heating of the refractory insulation layer. During the insulation stage, the voltage at both ends of the electrode is taken to 9 V for 12 h, and the temperature inside the furnace is maintained above 1700 ℃. At this point, the power supply is completed for 36 h, and the fluoride volatilization rate is 92.1%. From the temperature cloud map, it can be seen that the high-temperature zone during the insulation process no longer diffuses outward. In addition, the evolution of the cloud map reveals that the electrodes at both ends are a major part of heat dissipation, which is caused by the high thermal conductivity and specific heat capacity of graphite itself, while graphite felt and rock wool have good insulation performance in the upper part of the furnace type due to their small thermal conductivity.

To provide a detailed explanation of the temperature changes of the furnace material and refractory insulation layer at different positions during the heating process with heating time, a straight line at the cross-section of the high-temperature furnace center is selected as a reference, 640 mm from the bottom of the furnace, with a total length of 890 mm. The temperature changes are studied from the wall to the center of the furnace core. Fig. 2-38 shows the temperature distribution curve at different times on the horizontal line inside the furnace. Ⅰ to Ⅵ correspond to rock wool, insulation bricks, high alumina bricks, fillers, carbon bricks, and furnace materials in sequence (as the furnace core included is only 25 mm, it will not be listed here). From the figure, it can be seen that the heat during the heating stage is mainly used for the heating of furnace materials and carbon bricks, with the filler layer serving as the transition section and the insulation section having a relatively small temperature change; After the voltage increases to 9 V, the temperature of the furnace material decreases, and the heat increase of the rock wool, insulation brick layer, and high alumina brick layer significantly increases.

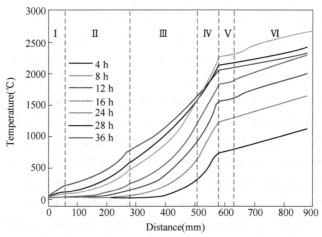

Fig. 2-38   Variation in temperature of monitoring line with changing of heating time

The temperature cloud map of five layers of refractory insulation materials along the monitoring line during the insulation process is shown in Fig. 2-39. From the figure, it can be seen that as the insulation time prolongs, the temperature change of the carbon brick layer is relatively small, and the high-temperature zone of the filler layer slightly decreases. The temperature of high aluminum bricks, insulation bricks, and rock wool all increase, especially in the insulation brick layer, where the temperature rise is more obvious. Based on Fig. 2-38, it can be observed that the temperature of carbon bricks and furnace waste cathode during the insulation stage is basically maintained at 1859 ℃. This indicates that the voltage setting in the insulation section is reasonable. At this stage, the voltage is mainly used for heating the insulation bricks and rock wool, as well as for heat dissipation on the furnace wall.

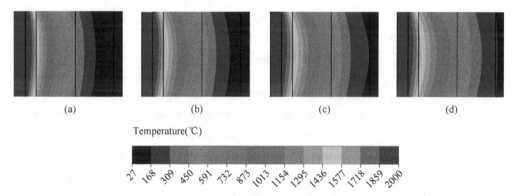

Fig. 2-39 Variations in temperature of refractory insulation along monitoring line at different heating time (X = 1.605 m)
(a) 4 h; (b) 8 h; (c) 12 h; (d) 16 h

Resistance furnaces generally have problems of high energy consumption and long cycles. Reasonable voltage distribution can effectively improve production efficiency, reduce product energy consumption, and thus save production costs. Loading voltage can cause changes in the internal heat source, thereby altering the distribution of the temperature field. Select the heating stage of the first 24 h as the research object to explore the influence of the pressure supply curve on the temperature distribution inside the high-temperature resistance furnace. To ensure that the heating capacity of the resistor material is basically the same, the power supply curve should meet the requirement of equal total power supply. The three types of voltage supply situations are shown in Fig. 2-40, which change every 8 h. The power supply voltages are decreasing (Scheme 1), constant voltage stable (Scheme 2), and increasing (Scheme 3), respectively. Study the uniformity of material area temperature and the proportion of temperature above 1700 ℃ through numerical simulation analysis of the lowest temperature change in the furnace, namely the volatilization rate of fluoride salts.

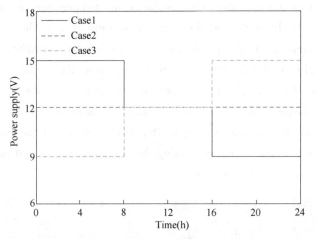

Fig. 2-40 Modes of power supply

Take the lowest point of the furnace temperature as the monitoring point, and the temperature changes at the monitoring point over time under the three power supply curves are shown in Fig. 2-41. The experiment measured that the fluoride in the waste cathode began to evaporate at 900 ℃ and completely evaporated when the temperature reached 1700 ℃. Observing the three curves in the graph, the temperature at the monitoring point approaches 900 ℃ after heating for 6 h when the voltage gradually decreases. The residence time above this temperature is as high as 18 h, which is about 9 h earlier than the time when the volatilization temperature is reached under the condition

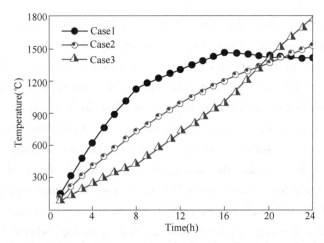

Fig. 2-41 Variations in Temperature with heating time at monitoring point

of gradually increasing voltage. The size of the waste cathode carbon block is 30-70 mm, so the fluoride salt needs to be treated under high temperature conditions for a certain period of time to fully evaporate, which is also the advantage of Scheme 1. From Fig. 2-42, it can be seen that when the power supply voltage decreases, the volatilization rate of fluoride salt reaches a peak of about 95% after heating for 16 h, meeting the production design requirements. After that, the volatilization proportion decreases, but it is basically maintained at 85%. The furnace temperature meets the requirement for fluoride salt to fully volatilize for up to 20 h, far exceeding the increasing situation. When the voltage increases, the sudden change in the proportion of temperature above 1700 ℃ inside the furnace lags behind, which is not conducive to the treatment of waste cathode carbon blocks in aluminum electrolysis cells. The change when the voltage is constant is between increasing and decreasing. In summary, the gradually decreasing pressure supply method is more advantageous for actual production, which is consistent with the research conclusion of relevant resistance furnace literature.

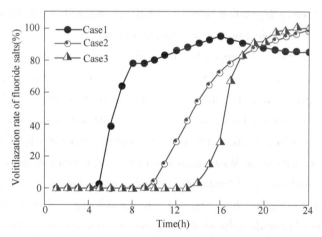

Fig. 2-42 **The volatilization rate of fluoride salts at different time**

This article uses the method of electrical thermal coupling numerical simulation to numerically calculate the separation process of carbon and fluoride salts in the waste cathode carbon block of aluminum electrolysis cells by electrically heating the electric heating high-temperature resistance furnace. The temperature field distribution and optimized power supply adjustment parameters of the electric heating high-temperature resistance furnace are obtained. The conclusion is as follows:

(1) The experiment shows that the effective removal temperature of fluoride from waste cathode carbon blocks needs to be ≥ 1700 ℃, and the volatilization rate of fluoride can reach over 93.1% at this temperature.

(2) The designed high-temperature resistance furnace radiates outward from the furnace core during the heating stage, with a maximum temperature of 2250 ℃ within the furnace after 24 h of heating. The effective volatilization area of fluoride salts accounts for 98%; During the insulation stage, the temperature of the waste cathode carbon block of the furnace material is basically maintained at around 2000 ℃. Under the power supply mode of 12 V heating for 24 h and 9 V insulation for 12 h, the processing time with an effective high temperature zone accounting for more than 80% can reach up to 18 h;

(3) Through optimization research on the power supply voltage mode, it was found that compared with the constant voltage and increasing voltage power supply modes, the effective heat treatment time in the furnace under the gradually decreasing power supply mode can reach 20 h, which is more conducive to the deep separation of fluoride salts.

2.5.4 Numerical simulation of the electro-thermal coupling treatment process

Aluminum spent cathode carbon block (SCCB) is defined as a hazardous material because it contains soluble fluoride and toxic cyanide compounds. To make the SCCB harmless, it is proposed to separate the carbon and fluoride salts through high-temperature resistance furnace. To evaluate the temperature distribution and heat transfer of furnace, effective thermal conductivity of SCCB was calculated by numerical simulation, and the results were compared with fundamental structural models. After considering radiative heat transfer, the effective thermal conductivity of SCCB increases as the temperature increases. When high-temperature resistance furnace was heated for 10 h under two conditions, the temperature considering radiation at monitoring point was found to be 61 ℃ higher than that without considering radiation. In addition, the volatilization rate of fluoride salts of the two models differed by 15.6%. This indicates that the temperature distribution in the furnace is uniform when radiation effect is included.

Aluminum is the second largest metal in the worldwide, and it is mainly produced using electrolytic cells. The cathode and refractory insulation materials in aluminum electrolytic cells continuously adsorb electrolytes and should be replaced after 5-8 a. When a cell cannot meet production needs, it is overhauled. It was reported that 30-50 kg spent cathode carbon block(SCCB) is discharged when one ton primary aluminum is produced. SCCB is considered as a hazardous material because it contains soluble fluoride and toxic cyanide compounds. The main valuable component of SCCB contains carbon and fluorine salt.

Most SCCBs are treated under storage because they contain soluble fluoride salts and cyanide, which are harmful to soil and water. Therefore, a detoxification treatment is required before storage, but the treatment cost is high. Other small-scale research methods currently include the coprocessing of SCCB. For example, as a fuel, SCCB is used in brick kiln plant and cement plant production. In the metallurgical field, SCCB can participate in ironmaking due to the presence of carbon, and the fluorine salts combined with limestone can be used as a flux calcium fluoride(CaF_2) in steelmaking. However, the sodium present in SCCB can cause damage to the lining of furnace. In previous studies, some high-temperature treatment methods were used for the detoxification of SCCB. Yao Wu et al. proposed a vacuum distillation method(VDM) to separate and recycle a SCCB, and the temperature and pressure were found to be important factors for the distillation ratio. The pyrometallurgy method is easy to implement, the amount of processing is large, and the degree of secondary pollution is small, making this suitable for the large-scale processing of SCCB. However, no studies have focused on the thermal conductivity of bulk material or the heat transfer mechanism in the furnace when a SCCB is treated at a high temperature.

Numerical simulations are used to study black box operation in high-temperature furnaces, and they play a vital role in the design and operation for metallurgical furnaces. Thermal conductivity is an important physical parameter for governing heat transfer. In a high-temperature resistance furnace, the furnace hearth with porous structures can be regarded as a mixing place for SCCB and air. Effective thermal conductivity is a key thermophysical property and plays an integral part in the design and continued performance of a furnace.

Many analytical models have been proposed to predict effective thermal conductivity. The fundamental structural models contain Series, Parallel, Maxwell-Eucken1 (ME1), Maxwell-Eucken2(ME2), and effective medium theory(EMT). Carson et al. organically combined the series and parallel methods by proposing a weighted average coefficient factor f, which is difficult to determine accurately. Lun et al. used Laplace equation to derive the effective thermal conductivity for multiphase materials with different structures. Wang et al. derived a Series+Maxwell-Eucken2 model in terms of volume fractions. However, the models of Series, Parallel, ME, EMT, and related models do not consider radiation when determining the effective thermal conductivity. Carbon and fluoride salts in the SCCB volatilize near 2000 ℃; thus, radiation heat transfer cannot be ignored.

In this study, numerical simulation is proposed to describe the thermal conductivity of SCCB in high-temperature resistance furnace during the heating. The results were fitted with the basic structural models. Furthermore, radiation heat transfer was considered in the calculation of effective thermal conductivity, and the effective thermal conductivity of SCCB was calculated at different temperatures. Meanwhile, the heat transfer characteristics of the influence on radiation were compared.

A model of a high-temperature resistance furnace, similar to a graphitization furnace, and an effective thermal conductivity model used for the simulations are shown in Fig. 2-43. The dimensions of the furnace are 3210 mm×1780 mm×1680 mm (length×width× height). The length of furnace hearth is 1600 mm; the depth and width of furnace hearth are 500 mm. The horizontal furnace also contains an electrode and five insulating layers, including carbon brick, insulating filler, high alumina brick, mullite, and rock wool. The electrode runs through the insulating wall and protrudes at both sides. The outer side is connected to a power supply, and the inner side is connected to the furnace core, as shown in Fig. 2-43(a).

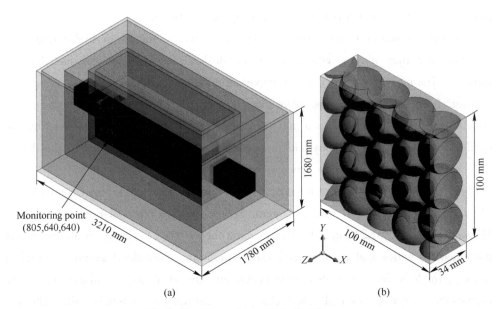

Fig. 2-43 Dimensions of studied furnace

(a) High-temperature resistance furnace; (b) Effective thermal conductivity model

One part of the furnace hearth was used to calculate effective thermal conductivity. To simplify the model, the diameter of a single-layer SCCB pellet was assumed to be 30 mm, and the pellets were arranged in a dense stack. The porosity in the fluidized bed was calculated to be 0.401 using the volume proportion of SCCB and gas. The model shown in Fig. 2-43(b) was used to calculate the steady-state heat transfer characteristics of pellets, and the effective thermal conductivity of SCCB was derived.

The simulation space contains fluid and solid, and a numerical simulation model was established. The gas and solid phases were described by its own mass, momentum, and energy equations. Considering that particles in the furnace are stationary, only the energy of gas and solid phases should be considered.

Effective thermal conductivity was determined from the thermal conductivity of flat wall combined with heat transfer and radiation equations. The heat of high-temperature furnace was provided electrically. Current was transmitted to the furnace core through the electrode, which heated the SCCB. The simulation of heat treatment furnace primarily considers the conduction, radiation, and convective heat transfer at the wall.

Heat is transferred via conduction and radiation. The partial differential equation describing heat conduction can be expressed using Eq. (2-32):

$$\rho c_p \frac{\partial T}{\partial t} = \frac{\partial}{\partial x}\left(\lambda_{\text{eff}} \frac{\partial T}{\partial x}\right) + \frac{\partial}{\partial y}\left(\lambda_{\text{eff}} \frac{\partial T}{\partial y}\right) + \frac{\partial}{\partial z}\left(\lambda_{\text{eff}} \frac{\partial T}{\partial z}\right) + q_s \quad (2\text{-}32)$$

where ρ is the density; c_p is the specific heat; λ_{eff} is the effective thermal conductivity; q_s is a heat source term; T is the absolute temperature; t is the time.

The physical parameters are shown in Table 2-12.

Table 2-12 Thermophysical properties used in this study

Parameters	Density (kg/m³)	Specific heat (J/(kg·K))	Conductivity (W/(m·K))	Conductivity (S/m)
Carbon block	1500	800	10	1×10^{-10}
Insulating filler	450	1465	0.65	1×10^{-10}
High alumina brick	1500	1100	0.786	1×10^{-10}
Insulating brick	1000	1200	0.3	1×10^{-10}
Rock wool	135	900	0.06	1×10^{-10}
Graphite felt	1400	900	0.01	1×10^{-10}
SCCB	850	840	—	2127
Electrode	1600	700	12.4	1.13×10^{-5}

Joule heat generated from an electric current was used to define the heat source term, which can be expressed as follows:

$$q_s = \tau |\nabla \varphi|^2 \quad (2\text{-}33)$$

where τ is the resistivity, and φ is the voltage.

The electrode and furnace core act as conductors, and the applied voltage obey the following differential equation:

$$\frac{\partial}{\partial x}\left(\gamma \frac{\partial \varphi}{\partial x}\right) + \frac{\partial}{\partial y}\left(\gamma \frac{\partial \varphi}{\partial y}\right) + \frac{\partial}{\partial z}\left(\gamma \frac{\partial \varphi}{\partial z}\right) = 0 \qquad (2\text{-}34)$$

where γ is the conductivity.

Convective heat transfer between the furnace sidewall and environment can be considered as follows:

$$-\lambda \frac{\partial T}{\partial n} = h_s(T - T_e) \qquad (2\text{-}35)$$

where λ is the thermal conductivity; T is the wall temperature; T_e is the ambient temperature; h_s is the heat transfer coefficient.

The heat transfer includes radiation heat transfer and convective heat transfer:

$$h_s = h_c + h_r \qquad (2\text{-}36)$$

where h_c is the convective heat transfer coefficient; h_r is the radiation heat transfer coefficient.

$$h_c = Nu(\lambda/L) \qquad (2\text{-}37)$$

where λ is the thermal conductivity; Nu is Nusselt number; L is the qualitative dimension of heat transfer direction.

$$h_r = \varepsilon\sigma(t_1^4 - t_0^4)/(t_1 - t_0) \qquad (2\text{-}38)$$

where ε is the radiation blackness; σ is the Stefan-Boltzmann constant, 5.67×10^{-8} W/(m$^2 \cdot$ K^4); t_1 is the wall temperature; t_0 is the ambient temperature.

The heat transfer of refractory insulation layer and materials in furnace follows the Fourier law of heat conduction.

$$q = -\lambda_i \frac{dt}{dx} = \lambda_i \frac{\Delta t_i}{\delta} \qquad (2\text{-}39)$$

where λ_i is the thermal conductivity of layer i; q is the heat conducted through insulation layer; δ is the thickness of layer.

The solution of effective thermal conductivity can be expressed as follows:

$$Q = -\lambda_{\text{eff}} \frac{t_{w1} - t_{w2}}{\delta} A \qquad (2\text{-}40)$$

where Q is the heat flow wall, W; δ is the thickness of model, m; A is the area of heat flow, m^2; t_{w1} is the temperature of wall 1, ℃; t_{w2} is the temperature of wall 2, ℃; λ_{eff} is the effective thermal conductivity that should be determined, W/(m · K).

Assuming that the temperature difference between the two walls is 100 ℃, the rest of the walls was set as adiabatic conditions. Effective thermal conductivity was derived from the wall heat flux.

The three-dimensional domain was calculated using the precision separation implicit algorithm in the numerical simulation. The prediction of heat transfer is usually quite sensitive to the number of grids, and it is important to use adequate cells while solving the governing equation. The total elements of the high-temperature resistance furnace and effective thermal conductivity model were 405 thousand and 103 thousand, respectively. The mesh and structure of furnace are shown in Fig. 2-44.

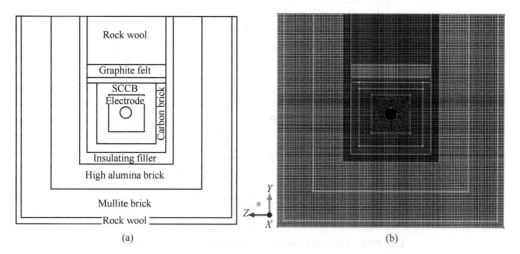

Fig. 2-44　Computational grid and structure of furnace
(a) Cross-section of high-temperature resistance furnace ($x = 1.605$ m);
(b) Mesh of section

The high-temperature resistance furnace consists of air and carbon block waste cathode. The thermal conductivity of air was 0.026 W/(m · K), while the thermal conductivity of carbon block waste cathode was determined to be 14 W/(m · K). The ME model has a good applicability when the content of dispersed phase is low, and the difference between two-phase thermal conductivity is small. ME1 model was found to be suitable for determining the thermal conductivity of continuous phase, which is greater than the dispersed phase, while the opposite result was obtained when ME2 was used[17]. Therefore, compared with ME1, the ME2 model is more suitable for the research needs of this study.

During actual production, the SCCB was piled up inside the furnace. The thermal conductivity of piled material has certain difficulty in the actual measurement. First, three fundamental structural thermal conductivity models (Parallel, Series, and ME2) were used to describe the porous medium; then, the results were compared when using numerical simulation without considering radiation. Because the SCCB contains fluoride that is harmful to the environment, the thermal conductivity in the laboratory was measured below 800 ℃. The thermal conductivity of a single SCCB at normal temperature (25 ℃), 200 ℃, 400 ℃, 600 ℃, and 800 ℃ was first measured experimentally, and the effective thermal conductivities of Parallel, Series, ME2 models were calculated, as well as the numerical results, as shown in Fig. 2-45.

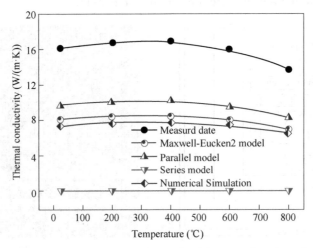

Fig. 2-45　Effective thermal conductivity as a function of temperature calculated using different models

Without considering radiation, the effective thermal conductivity calculated using the empirical formulas have the same trend as that determined from numerical simulation, and the thermal conductivity first increased with temperature and then decreased (Fig. 2-45). The thermal conductivity is 6.51 W/(m · K) when heated to 800 ℃. Compared with the results reported in the literature, graphite materials exhibit a similar trend with temperature. Because the content of graphite in the SCCB is 60%, a certain amount of fluoride salt and cyanide exists in the middle. The thermal conductivity of air is small; therefore, the effective thermal conductivity of furnace hearth is lower than that reported in the literature.

The Parallel model for effective thermal conductivity calculated using different models has the highest corresponding value at different temperatures, while the Series model has the lowest effective thermal conductivity. This is because in the Parallel model, heat flow and superposition of the two substances are parallel, while in the Series model, heat flows perpendicular to the direction in which the two substances are superposed. The effective thermal conductivity of a general composite material is defined in Series and Parallel model. The value of effective thermal conductivity obtained from numerical simulation is similar to that determined using the ME2 model, and the values obtained lie between those determined from the Parallel and Series models. The ME2 model was derived using the Laplace equation, and the numerical simulation results were calculated using finite-difference time-domain simulations. The two solution modes have similar characteristics and provide an approximate solution of differential equation. The exact solution of effective thermal conductivity of porous media is slightly lower than that of ME2 model. These results show that it is feasible to calculate the effective thermal conductivity of porous carbon block waste cathode in a resistance furnace using numerical simulation.

The SCCB in high-temperature furnace was heated at normal pressure, and the temperature should reach 1700 ℃ to completely separate the fluoride from the carbon. Radiative heat transfer between the materials plays a crucial role at high temperatures. The effective thermal conductivity of SCCB deposit is shown in Fig. 2-46. The effective thermal conductivity increased as the temperature increased. When the temperature was below 1000 ℃, the effective thermal conductivity changed slowly with temperature. The thermal conductivity at 800 ℃ was 8.09 W/(m·K), whereas it was 16.08 W/(m·K) at 1800 ℃. A comparison of Fig. 2-45 and Fig. 2-46 shows that thermal conductivity decreases with increasing temperature when radiation is not considered. When the isothermal surface moves in a single layer of material sphere, heat is primarily transferred via conduction; when the isothermal surface is transmitted across the material layer, heat is transferred via conduction and radiation between balls.

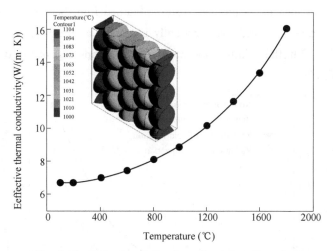

Fig. 2-46 Effective thermal conductivity of stacked material as a function of temperature

A self-designed high-temperature resistance furnace was used to treat a SCCB from an reduction cell to recover the carbon material and fluoride salts and remove the hazardous cyanides. The area in the furnace above 1700 ℃ was defined as the effective volatile area, and the ratio of volume above 1700 ℃ to the total volume was defined as the evaporation rate of fluoride salt. The volume fraction for different temperature intervals was obtained by integrating the temperature specified in the selected area of model by the function in CFD-Post.

To explain the effect of radiation on the results, the two effective thermal conductivities of calculated by numerical simulation were simulated. At present, the size of core in small industrial furnaces is 100 mm×100 mm×1600 mm. When the voltage is 9 V, the corresponding current is 5000 A, the corresponding furnace core resistivity is 8.84 μΩ · m, and the furnace power is 45 kW. The temperature over time at the monitoring point is shown in Fig. 2-43, and the volatilization rate of fluoride salt after 10 h is shown in Fig. 2-47. At the beginning of heating, the temperature at the monitoring point slowly increases. After heating for 4 h, the calculated temperature increase rate at the monitoring point was larger when radiation was considered than that when radiation was not considered. The temperatures when radiation was considered or not at the monitoring point under heating for 10 h were 1092 ℃ and 1031 ℃, respectively. Further, before reaching 800 ℃, the calculated effective thermal conductivity including radiation is smaller than that calculated without radiation. Radiative heat transfer becomes larger above 1000 ℃, and more heat is transferred outwards from the furnace core, causing the monitoring point to heat up faster. The results obtained from both the models show that the volatilization rate of fluoride salt gradually increased as the heating time increased. After heating for 10 h, the volatilization rates of the fluoride salt of the two models were 41.1% and 56.7%.

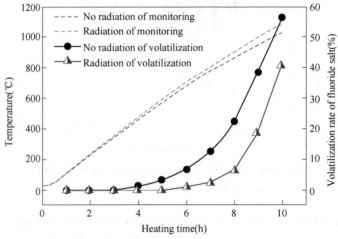

Fig. 2-47 Temperature at monitoring point and volatilization rate of fluoride at furnace in two calculated thermal conductivities

Fig. 2-48 shows the temperature distribution in the two models after heating for 10 h. The temperature distribution determined with the two thermal conductivities calculated had similar trends, and heat transfers from the center to outwards. Fig. 2-48(a) shows that the heat generated by the core is more concentrated in the center of furnace when radiation is not considered. When effective thermal conductivity is calculated with radiation, as shown in Fig. 2-48(b), more heat diffuses from the furnace core to the surroundings, and the temperature of lining material increases, corresponding to the temperature trend at the monitoring point shown in Fig. 2-47. Meanwhile, it is consistent

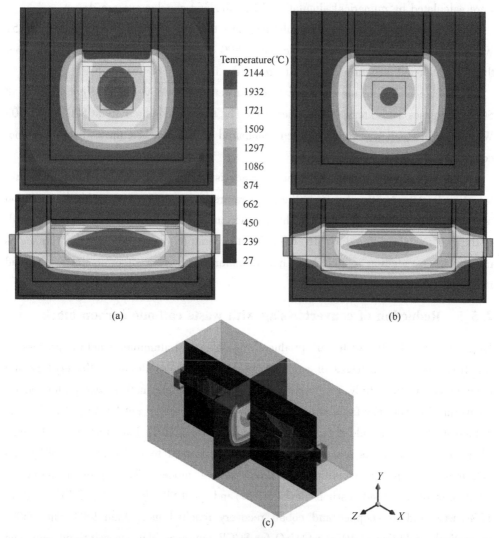

Fig. 2-48 Temperature field after heating for 10 h ($X=1.605$ m and $Z=0.89$ m)

(a) No radiation; (b) Radiation; (c) Location

with the lower evaporation rate of fluoride salt when radiation is considered. Fig. 2-47 and Fig. 2-48 show that the temperature distribution in a furnace is more uniform when radiation is included. This shows that radiation has an irreplaceable role in the simulation of heat transfer in a high-temperature furnace for treating SCCB.

The SCCB obtained from aluminum reduction cells is considered as a hazardous material. A pyrometallurgy method is proposed to effectively separate the carbon and fluoride salts in SCCB. Herein, a self-designed high-temperature resistance furnace is proposed, and the effective thermal conductivity of SCCB and heat transfer characteristics were calculated by numerical simulation. The following conclusions are drawn:

(1) The results of numerical simulation show that the thermal conductivity is 6.51 W/(m·K) when the furnace was heated to 800 ℃, consistent with the results of other three fundamental structure models. It proves the feasibility of this method.

(2) After considering the radiative heat transfer, the effective thermal conductivity of SCCB increases as the temperature increases. The thermal conductivity at 800 ℃ is 8.09 W/(m·K), which was quite different compared with that when the radiation was not considered.

(3) As a high-temperature resistance furnace is heated, heat radiates from the center to outwards. After heating for 10 h, the temperature at monitoring point with the radiation model was 61 ℃ higher than that without radiation. The volatilization rate of fluoride salt of the two models differed by 15.6%. The temperature distribution in the furnace is uniform when radiation is included. Thus, radiation has a great effect on the results, and it cannot be ignored.

2.5.5 Reduction of converter slag with waste cathode carbon block

Large amounts of solid wastes are produced in copper and aluminum smelting processes, which not only causes losses of valuable resources, but also threatens the ecology and environment. In this study, a reduction-sulfurization smelting method was performed for recovering Cu and Co from converter slags by using spent pot-lining (SPL) as the reductant. CaO was added to fix the fluorine from SPL into the cleaned slag. Thermodynamic analysis and experiments were performed to verify the feasibility and determine the optimal conditions of this smelting process. The optimum reductant addition of spent cathode carbon block (SCCB) and spent SiC side block (SSCB) was 8%-12% (wt), and the copper and cobalt recovery reached more than 98% and 96%, respectively. Addition of 10% (wt) CaO for SCCB improved slag viscosity and promoted the separation between slag and matte/alloy as well as fixed fluorine in the cleaned slag in the form of insoluble calcium fluoride. The metallized Cu-Co matte was obtained, in which Cu mainly existed in the form of sulfide and Co mainly existed in the form of iron cobalt alloy.

In the extraction of non-ferrous metals, large amounts of wastes are discharged during the smelting process and at the end of the service life of smelting equipment. The environmental issues arising from solid wastes such as slag, ash, tailings and spent potlining have become increasingly worthy of attention[66-69]. Copper and aluminum are the two most produced non-ferrous metals in the world and the solid wastes generated during the extraction of these metals amount to millions of tonnes every year.

Fig. 2-49 shows the productions of smelted copper and primary aluminum and the generations of solid wastes of copper slag and SPL. The total world copper smelting production tonnage reached about 20 Mt/a in the recent 5 years. It is estimated that about 2.2 t of slag are generated for every ton of copper production[70], and thus more than 40 Mt copper slag is discharged per year. For primary aluminum, the annual

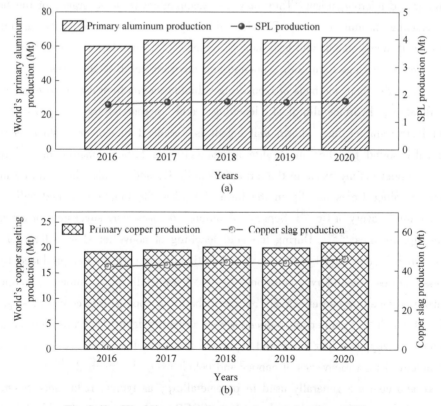

Fig. 2-49 World's production of primary aluminum, SPL, smelted copper and copper slag in the last 5 years

output has reached about 60 Mt/a in the recent 5 years. The typical service life of the modern reduction cell is about 2500-3000 d. About 24-30 kg of spent pot-lining (SPL) are generated for every ton of primary aluminum and about 1.7 Mt of SPL are discharged per year[71]. Although the copper slag and SPL are solid wastes, they contain many valuable elements, such as copper, cobalt, and carbon. Therefore, the key issue for processing these solid wastes is effective harmless treatment and economic recycling.

The converter slag which has high recycling values usually contains high grade of valuable metals, such as copper, cobalt, nickel and zinc[71-72]. Increasing the recovery ratio of these valuable elements will reduce their circulations in the smelting system and increase the product yield and added value, which will make a high economic value. On the other hand, the content of heavy metals in the untreated converter slag is high, which has potential harm to environment[73]. Studies have shown that heavy metals such as copper and cobalt in converter slags can seep into groundwater under the actions of rainwater and microorganisms. They may also accumulate in some organs of the human body, causing chronic poisoning. Proteins and enzymes in the human body can interact strongly with heavy metals and lose their activities[74-75]. Therefore, it is essential to treat copper converter slags from both economic and environmental perspectives.

Currently, the main methods of recovering valuable metals from copper converter slags are flotation, hydrometallurgical leaching, and pyrometallurgical processes[76]. Mineralogical studies have shown that cobalt is dissolved in converter slags as oxides chemically bound to silica in fayalite or magnetite as lattice substituents[77]. In the tailings, copper mainly exists in the form of metallic Cu and sulfide which can be floated effectively. Since Co is usually in the form of oxides, the flotation method will not be effective for recovery of Co[78]. Separate flotation processes were suggested for the mixed oxide-sulphide ores[79-80]. During the acid leaching of converter slags, silica gel often forms, which hinders metal extraction[81]. Additionally, the wastewater produced by this process causes secondary pollution. As a pyrometallurgical process, reduction-sulfurization smelting can effectively recover cobalt and copper from converter slags. The oxides of cobalt and copper are reduced with a reducing agent. They are then sulfurized to produce a copper-cobalt matte and separated from the cleaned slag. Some researchers have achieved high recoveries of copper and cobalt using this method[82-83].

Coke and coal are generally used in pyrometallurgy as typical reductants. Spent pot-lining including spent cathode carbon block (SCCB) and spent SiC side block (SSCB) which are generated at the end of service life the aluminum reduction cell (ARC) can also be used as a reductant during the high temperature processing[84-85]. During the service life of ARC, the cathode and side linings are constantly eroded and saturated, causing the linings expansion and deterioration. Thus, considerable quantities of harmful and toxic components are adsorbed in SPL[86-88]. The efficient utilization and recycling of SPL has become an urgent problem in the aluminum industry. The fluoride in SPL can be completely removed by hydro-processing, and high temperature > 1973 K or high vacuum is required for pyro-processing, which has high requirement for the equipment[89-90].

SPL is a hazardous solid waste produced by aluminum smelting industry, and no economic and reasonable treatment and utilization technologies have been widely used in industry. The existence of fluoride limits the recovery and further utilization of SPL. At present, many attentions are focused on the pyrometallurgical treatment of SPL. However, the equipment and economy have not meet the requirements of industrial production. Aluminum and copper are the two nonferrous metals with the largest output and their extraction technologies are fully developed. This work proposes a route to achieve collaboration between aluminum and copper industries. The treatment and utilization of SPL in aluminum industry was combined with the smelting process of copper converter slag. This method avoids the problem of defluorination of SPL, that is, carbon and silicon carbide are used, and the toxic components (soluble fluoride and cyanide) are transformed and removed during the smelting process. Cyanide decomposes completely at high temperature and more than 90% (wt) of fluorine will be transformed into calcium fluoride and fixes in the slag. Because of the high content of magnetic iron in converter slag, the viscosity of slag is high. The generated CaF_2 can improve the fluidity of slag and improve the separation of slag and matte. The sodium can be collected as dust in the form of oxide, and fluoride can be removed along with tail gas desulfurization process. It is to be expected that using SPL as reductant for smelting process of copper converter slag may not only realize the harmless treatment and utilization of SPL, but also has good economy and feasibility. However, the potential impacts of adding SPL as reductant in converter slag smelting process need to be considered carefully.

Principle, Apparatus and Procedure. The principles for RSS of converter slags are as follows:

(1) In the reducing process, the oxidized iron, copper and cobalt are reduced into low valent oxides and metals. The carbon in SCCB or SiC in SSCB is used as reductant. A part of Fe_3O_4 is reduced into FeO, and a part of FeO, Cu_2O and CoO are reduced to

metallic Fe, Cu and Co, respectively, at a high smelting temperature of >1573 K.

(2) In the sulfurizing process, most of Cu_2O are sulfurized to form a matte phase of $mCu_2S \cdot nFeS$ and $m'CoS \cdot n'FeS$ by smelting with a sulfurizing agent such as chalcopyrite concentrate. A part of Fe_3O_4 reacts with FeS and SiO_2 to form the slag phase of forsterite.

(3) The original metals of Cu and Co in the slag phase are captured and dissolved by the matte phase.

Thus, the Co-Fe alloy, Cu metal and Cu_2S can be recycled in the matte phase which is separated from the slag phase by density difference. The Cu and Co can further be separated by crushing, magnetic separation and leaching processes.

The reduction-sulfurization smelting experiments for extracting valuable metals from copper converter slags were performed in a 12 kV·A muffle furnace (Tianjin Zhonghuan Furnace Corp.). The temperature with a precision of ±1 K was monitored through a sheathed thermocouple inserted into the muffle furnace. 100 g of copper converter slag and 40 g sulfurizing agent chalcopyrite concentrate, the composition of which is shown in Table 2-13, were used for each test with a given amount of reductant of SCCB or SSCB. In addition, 10% (wt) of the CaO was added together with SCCB to fix the fluorine in the slag. The addition of SCCB, SSCB or CaO is defined as mass ratio of the added material to the converter slag.

Table 2-13 Chemical compositions of converter slag, sulfurizing agent, SCCB and SSCB (wt%)

Converter slag		Cu	Fe_2SiO_4	Fe_3O_4	Cu_2S	CoO	Al_2O_3	CaO	MgO	SiO_2
		8.56	47.6	32.1	1.24	2.31	1.01	0.09	0.15	6.94
Sulfurizing agent		$CuFeS_2$	FeS_2	SiO_2	Al_2O_3	MgO	CaO	CoO		
		76.7	15.6	5.47	1.46	0.56	0.18	0.05		
SPL	SCCB	C	NaF	SiO_2	AlF_3	Fe_2O_3	CaF_2	CN^-		
		68.1	20.1	3.95	6.17	0.73	0.98	0.002		
	SSCB	SiC	Si_3N_4	NaF	CaF_2	AlF_3	Fe_2O_3	SiO_2	CN^-	
		76.7	10.2	5.23	0.57	0.35	2.3	4.64	ND	

All raw materials were crushed and ground to finer than 74 μm by a vibration mill prototype (Wuhan Exploring Machinery Factory, China) before the experiments. The raw materials were uniformly mixed and transferred into a φ50 mm × H120 mm corundum crucible placed inside the muffle furnace. According to the previous studies, the optimal range of smelting temperature was from 1623 K to 1723 K, at which high slag matte separation rate, as well as high Cu and Co recover ratio, can be obtained[91]. Therefore, the smelting temperature was set as 1673 K in this work. The raw materials were heated from a room temperature to 1673 K at a rate of 10 K per minute and adequately reacted for 2 h. After the reduction and sulfurization reactions, the slag layer was floated above the matte phase. The products in the crucible were naturally cooled to the room temperature and the slag and the matte were separated and weighed. The experimental setup is shown in Fig. 2-50.

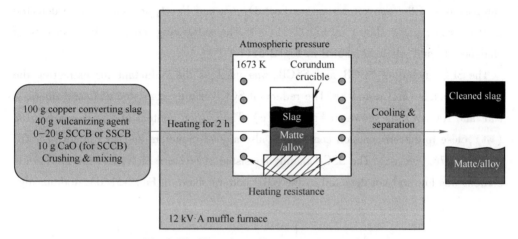

Fig. 2-50 Experimental setup and procedure

Analysis Method. The metal element contents in the raw materials, cleaned slag and matte were analyzed by inductively coupled plasma atomic emission spectrometry (ICP-OES, PerkinElmer, the USA). Aqua regia (hydrochloric acid: nitric acid = 3 : 1) was used to dissolve the tested sample. The solution was diluted 100 times in a 1 L volumetric flask, and then 10 mL was taken for testing. The recovery ratios of Cu and Co were calculated based on the mass balance principle. The phase compositions and microstructures of raw materials, the cleaned slag and the copper-cobalt matte were characterized by X-ray diffraction (XRD, Philips, The Netherlands), scanning electron microscope (SEM, JOEL, Japan), respectively.

100 g of the tested solid sample (SCCB, SSCB and the cleaned slag) was dissolved in 1 L of water for 1 to prepare the standard leaching solution for measuring solubilized F^- and CN^- contents. The solubilized F^- content was determined potentiometrically using a fluorine ion-selective electrode (ISE, PXSJ-216, Shanghai INESA Scientific Instrument, China) that had a minimum detection concentration of 0.05 mg/L. and the cyanide content was measured by a silver nitrate titration method (SEPA), which provided a minimum detection concentration of 0.025 mg/L[92].

Based on the theory of multiphase equilibrium, thermodynamic calculation on standard Gibbs free energies of the reactions, metal recovery ratio and slag viscosity were performed by FactSage 7.0.

Materials. The chemical and phase compositions of the experimental materials of the copper converter slag, sulfurizing agent and reductant are presented in Table 2-13 and Fig. 2-51, respectively. A converter slag with 1.82% (wt) Co and 9.85% (wt) Cu and a sulfurizing agent, chalcopyrite concentrate, are used in this study. The main phase compositions of the converter slag are Fe_2SiO_4, Cu and Fe_3O_4 and no Co were detected in the XRD pattern shown in Fig. 2-51(a). The sulfurizing agent mainly consists of chalcopyrite and silica, as shown in Fig. 2-51(b).

The SPL, including SCCB and SSCB, was used as the reductant for extracting the valuable metals Cu, Co and Fe. The reductant SCCB mainly consisted of C and fluorides NaF and Na_3AlF_6, as shown in Fig. 2-51(c). The carbon content of SCCB was 68.1% (wt), close to that in common coal. SiC is also a good reductant and its content in SSCB was up to 76.7% (wt). The SSCB mainly consisted of SiC and Si_3N_4 and the its fluoride content was low and not detected in the XRD pattern shown in Fig. 2-51(d). Additionally,

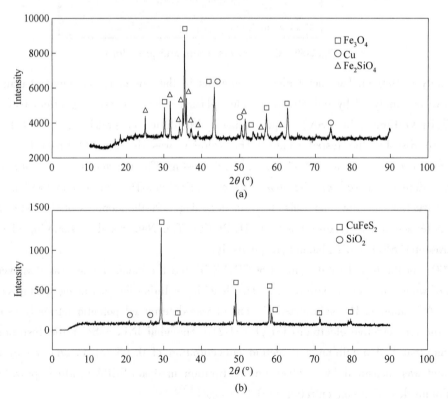

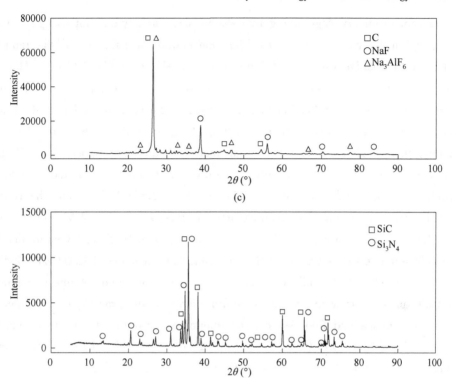

Fig. 2-51 X-ray diffraction pattern
(a) Copper converter slag; (b) Sulfurizing agent; (c) SCCB; (d) SSCB

the fluorides in SCCB and SSCB decrease the viscosity of the molten slag, accelerating the separation of metal/matte and slag and increasing the recovery of copper and cobalt. Although, the fluorides could improve the smelting process, they should be transformed into an insoluble phase to avoid their release into the atmosphere. Therefore, analytically pure CaO was used to prevent volatilization of the fluorides in the SPL, especially, NaF and AlF_3 in SCCB.

Thermodynamic calculations. The reaction mechanism of reduction-sulfurization smelting of the converter slag with SPL and sulfurizing agent is shown in Table 2-14 and the standard Gibbs free energies of the reactions are plotted in Fig. 2-52 at 1473 K to 1873 K and under an atmospheric pressure. SPL was used to control the oxygen potential of the molten slag and reduce the copper and cobalt oxides. In the reduction process, copper, cobalt and iron are reduced and extracted from the slag with SCCB and SSCB, and the possible reduction reactions are shown by reactions Eq. (2-44) and Eq. (2-48). Copper and cobalt can also be reduced by iron, as shown by reactions Eq. (2-49) – Eq. (2-50). The reduction of Fe_3O_4, shown by reactions Eq. (2-44) and Eq. (2-45), is very important for secondary slag-making, by reducing slag viscosity and improving separation of slag and matte/metals. At the smelting temperature of 1673 K, all the reduction reactions with negative value of $\Delta G^\ominus$ are favored as shown in Fig. 2-52(a). The $\Delta G^\ominus$ values of reactions of Cu_2O and CoO reduced by SPL are both lower than those for reduction by iron metal, indicating that the former is the main reduction mode. Copper oxide is more easily reduced than cobalt oxide from a thermodynamic viewpoint, indicating that cobalt recovery is somewhat more difficult than copper. The sulfurizing agent acts as a collector phase to provide a sufficient volume of matte and promote the separation of matte and slag. Reactions Eq. (2-51) – Eq. (2-53) show the possible sulfurizing reactions during the smelting process. As shown in Fig. 2-52(b), the $\Delta G^\ominus$ values for reactions (Eq. (2-51) – Eq. (2-53)) are negative 1673 K, indicating that FeS can easily sulfurize Cu_2O/Cu to Cu_2S, while sulfurizing reaction of CoO (Eq. (2-52)) can't be occurred 1673 K, indicating CoO easier to transform into metallic phase rather than matte phase. During the reduction-sulfurization smelting process, toxic components of soluble cyanides and fluorides can be eliminated according to possible reduction reactions (Eq. (2-54) – Eq. (2-57)). According to Eq. (2-54) and Eq. (2-56), water-insoluble CaF_2 may be formed by adding CaO to fix NaF and AlF_3 in the slag. The CaF_2 can decrease melting point and improve fluidity of the slag that may accelerate the separating rate of slag and matte/metal and promote the recovery of copper and cobalt. $CaO \cdot SiO_2$ cannot be generated (Eq. (2-55)) at the smelting temperature. Reaction (Eq. (2-57)) shows that cyanide can decompose easily to form CO_2 and N_2 at high temperature which eliminates the toxicity of SCCB.

Table 2-14 Reaction mechanism of reduction-sulfurization smelting process[93] (1473–1873 K)

	Reactions	$\Delta G_T^\ominus$ (J/mol)	No.
	$Fe_3O_4(l)+C(s)=3FeO(l)+CO(g)$	$\Delta G^\ominus = 240640-228.98T$	Eq. (2-41)
	$FeO(l)+C(s)=[Fe]+CO(g)$	$\Delta G^\ominus = 126585-135.78T$	Eq. (2-42)
	$CoO(l)+C(s)=[Co]+CO(g)$	$\Delta G^\ominus = 118897-155.91T$	Eq. (2-43)
	$Cu_2O(l)+C(s)=2[Cu]+CO(g)$	$\Delta G^\ominus = 6305-128.87T$	Eq. (2-44)
Reduction reactions	$3Fe_3O_4+SiC(s)=9(FeO)+CO(g)+SiO_2(l)$	$\Delta G^\ominus = 130569-356.36T$	Eq. (2-45)
	$3FeO(l)+SiC(s)=3[Fe]+CO(g)+SiO_2(l)$	$\Delta G^\ominus = -211596-76.77T$	Eq. (2-46)
	$3CoO(l)+SiC(s)=3[Co]+CO(g)+SiO_2(l)$	$\Delta G^\ominus = -234662-137.15T$	Eq. (2-47)
	$3Cu_2O(l)+SiC(s)=6[Cu]+CO(g)+SiO_2(l)$	$\Delta G^\ominus = -527436-56.03T$	Eq. (2-48)
	$Cu_2O(l)+[Fe]=2[Cu]+FeO(l)$	$\Delta G^\ominus = -109270-0.63T$	Eq. (2-49)
	$CoO(l)+[Fe]=[Co]+FeO(l)$	$\Delta G^\ominus = -7689-20.13T$	Eq. (2-50)

Continued Table 2-14

	Reactions	$\Delta G_T^\ominus$ (J/mol)	No.
Sulfurizing reactions	$Cu_2O(l) + [FeS] = [Cu_2S] + FeO(l)$	$\Delta G^\ominus = -150880 + 19.76T$	Eq. (2-51)
	$CoO(l) + [FeS] = [CoS] + FeO(l)$	$\Delta G^\ominus = -39832 + 30.30T$	Eq. (2-52)
	$2[Cu] + [FeS] = [Cu_2S] + [Fe]$	$\Delta G^\ominus = -30600 + 12.85T$	Eq. (2-53)
Detoxification reactions	$NaF(l) + 0.5CaO(l) + 0.5SiO_2(l) = 0.5CaF_2(l) + 0.5Na_2O \cdot SiO_2(l)$	$\Delta G^\ominus = -63477 + 15.02T$	Eq. (2-54)
	$NaF(l) + 1.5CaO(l) + SiO_2(l) = 0.5CaF_2(l) + 0.5Na_2O \cdot SiO_2(l) + CaO \cdot SiO_2(l)$	$\Delta G^\ominus = -555343 - 64.52T$	Eq. (2-55)
	$AlF_3(l) + 1.5CaO(l) = 1.5CaF_2 + 0.5Al_2O_3$	$\Delta G^\ominus = -396054 + 100.56T$	Eq. (2-56)
	$NaCN(l) + 1.25O_2(g) = CO_2(g) + 0.5N_2(g) + 0.5Na_2O(g)$	$\Delta G^\ominus = -500091 + 28.68T$	Eq. (2-57)

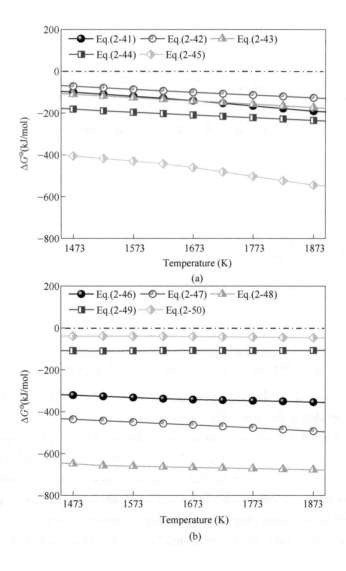

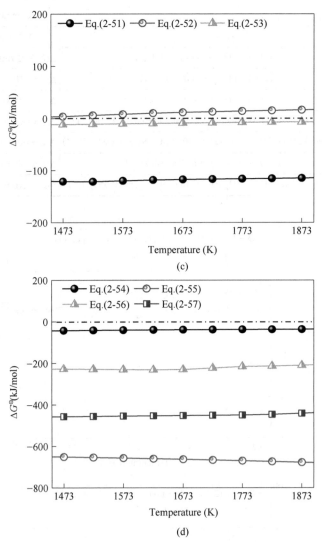

Fig. 2-52 Standard Gibbs free energies
(a)(b) Reduction reactions; (c) Sulfurization; (d) Detoxification reactions

The composition of in the cleaned slag was investigated based on thermodynamics equilibrium calculation by using FactSage software. Fig. 2-53(a) shows the amount of cleaned slag formed for different reductant additions. The main components in the cleaned slag were Fe_2SiO_4, CaO and Na_2O, which have great influences on slag property. Fig. 2-53(b)-(d) show the SiO_2, CaO and Na_2O contents in the cleaned slag for different reductant additions. As shown in Figs. 2-53(a) and (b), with an increase in SCCB and SSCB, the reduction degree of slag increased and the SiO_2 content also increased, which increased the slag viscosity. Further, the total amount of slag decreased by the iron content being sulfurized. Compare to SCCB, additional silicon was introduced into the slag when SSCB was used as the reductant. As shown in Fig. 2-53(c), the CaO content in the cleaned slag is very low without CaO addition. When adding 10% (wt) CaO, the CaO content decreased slightly at first and then increased rapidly with increasing reductant addition, mainly because of the variation of the cleaned slag amount. Na_2O in the cleaned slag mainly came from the NaF in SCCB and it increased rapidly with an increase in SCCB addition. Due to low NaF content in SSCB, the Na_2O content changed slightly with increasing SSCB addition, shown in Fig. 2-53(d). As shown in Fig. 2-53(e), when adding 10% (wt) CaO, the fluorine was left in the form of CaF_2 in the cleaned slag, with the CaF_2 content increasing with increasing SCCB addition. It is noted that the CaF_2 content strongly decreases the cleaned slag viscosity.

According to the compositions of the cleaned slag with different reductant addition, its viscosity was calculated and shown in Fig. 2-53(f). The viscosities were calculated based on the viscosity database in FactSage and a modified quasichemical model[94-95]. The results show that the viscosity of the cleaned slag increased with increasing reductant addition, mainly because the increasing SiO_2 content. The higher the reduction degree of the cleaned slag, the higher its viscosity. When the reductant addition exceeded 12% (without CaO), the viscosity increased significantly, notably reaching more than 2 Pa · s with 14% (wt) SCCB. Therefore, the amount of reducing agent should be controlled to ≤ 12% (wt). Even then, the viscosity with 12% (wt) addition was still high. The addition of CaO reduced the viscosity due to the participation of CaF_2. From the composition and viscosity of the slag, SCCB has a better reduction effect than SSCB and adding CaO avoids the volatilization of fluorides and improves the constitution of the cleaned slag.

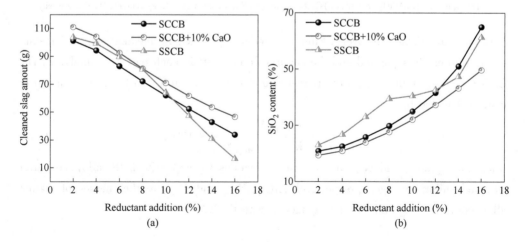

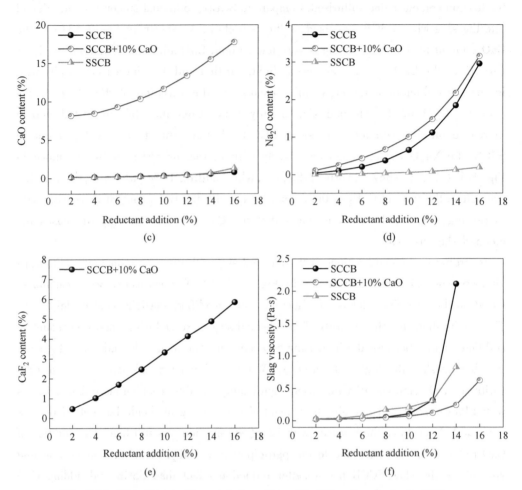

Fig. 2-53 Slag Composition and viscosities with different reductant additions
(a) Slag amount; (b) SiO_2 content; (c) CaO content; (d) Na_2O content; (e) CaF_2 content; (f) Slag viscosity

Copper and Cobalt Recovery. After the smelting process, the cleaned slag and metal/matte were easily separated and the Cu and Co contents in each part were analyzed. The Cu/Co recovery ratio, γ_{Me}, is the percentage of recovered Cu/Co in the form of metal or matte calculated by

$$\gamma_{Me} = \frac{c_{Me,ma} M_{ma}}{c_{Me,s} M_{slag} + c_{Me,v} M_{va}} \times 100\% \qquad (2\text{-}58)$$

where $c_{Me,ma}$, $c_{Me,s}$ and $c_{Me,v}$ are the mass fractions Cu or Co in matte/alloy, converter slag and sulfurizing agent, respectively; M_{ma}, M_{slag} and M_{va} are the masses of matte/alloy, converter slag and sulfurizing agent, respectively.

The Cu and Co recovery ratios are also calculated by FactSage 7.0 based on the Gibbs free energy minimization principle. The Cu and Co recovery ratios were compared between experimental and calculated results after a smelting at 1673 K and the results were exhibited in Fig. 2-54.

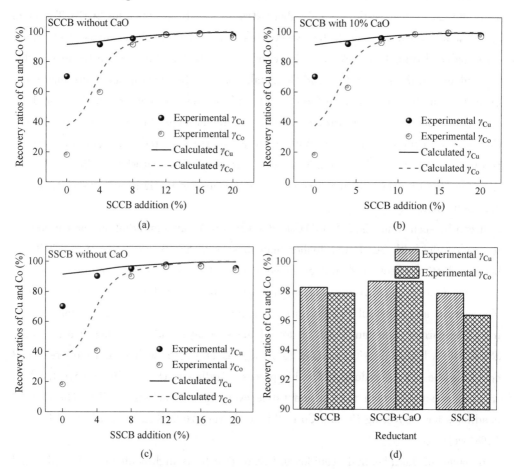

Fig. 2-54 Calculated and experimental recovery ratios of Cu and Co in form of matte/alloy, for different reductant additions (wt)
(a) SCCB; (b) SCCB+10% (wt) CaO; (c) SSCB; (d) Experimental recovery ratios of Cu and Co at a 12% (wt) reductant addition

Without reductant addition, only 70.2% (wt) of Cu and 18.2% (wt) of Co could be recovered from the raw materials in the experiments. Large parts of Cu and Co were left in the cleaned slag as Cu_2O and CoO, respectively. The experimental results were much lower than the calculated values, because: (1) Cu_2O and CoO were not effectively reduced without a reductant, and Co tends to combine with iron compound such as fayalite and iron oxide and thus is not transferred to the matte/alloy phase; (2) The sulfurizing reactions between Cu_2O/CoO and FeS are slow; (3) The high viscosity of the slag with a high Fe_3O_4 content impedes separation between slag and matte.

When adding 4% (wt) reductant, the copper recovery ratio reached more than 90%, while the Co recovery ratio was only about 40% for SSCB and 60% for SCCB. The main reason was that Cu is easier to be reduced than Co. When the reductant is insufficient, Cu is reduced prior to Co. With an increase in reductant addition, more Co in the slag was reduced in the form of CoS and Co metal. CoS was dissolved in the matte phase and Co metal combined with iron to form an alloy phase. With a reductant addition of 12% (wt), the reduction and sulfurization reactions were sufficient, the recovery ratios of both Cu and Co reaching about 98%. With a further increase in reductant addition, the Cu and Co recovery ratios changed only slightly. It is noteworthy that when adding 20% (wt) reductant, the experimental recovery decreased because of the large slag viscosity due to over reduction.

It can be seen from Fig. 2-54(d) that at a 12% (wt) reductant addition, the Cu and Co recovery ratios with SCCB was higher than those with SSCB, indicating that SCCB has better reduction effect. It should be noted that in the experimental conditions, the raw materials including converter slag, sulfurizing agent and reductant were well mixed. However, in an industrial process, the reductant is usually added from a top feeding inlet. The carbon reductant floats on top of the slag layer due to its low density, while SSCB with a higher density is easier to settle, mix and react with slag. Therefore, SSCB could also be a good reductant for this process. Adding CaO with SCCB gave the best smelting effect with both the Cu and Co recovery ratios reaching 98.7%. The optimal conditions for recovering Cu and Co in this work were 1673 K, 2 h, 12% (wt) SCCB and 10% (wt) CaO.

Behavior of fluorides and cyanides in SCCB. Due to its high contents of fluorides and cyanides, SCCB is considered as a hazardous solid waste. Therefore, when using SCCB as a reductant for extracting valuable metals from the converter slag, the behavior of the toxic components fluorides and cyanides should be considered carefully. The fluorides and cyanides in SCCB would decompose and volatilize during the smelting process, and parts of them would be transferred to the slag phase. The cyanide is easy to totally decompose by high temperature processing at 1673 K. However, large amounts of fluorides, such as Na_3AlF_6, AlF_3 and NaF, will be discharged into the air in the form of gases or remain in the slag in the form of soluble fluorides, which may cause secondary environmental pollution problems. A key point considered in this work was to avoid the fluorides and cyanides transferred to air, eliminate their toxicity and form insoluble phases that remain in the slag.

The toxicity of F^- and CN^- in the cleaned slag is analyzed for different SCCB additions shown in Table 2-15. The limits of the contents of soluble fluorides and cyanides should meet the requirements that $c_{F^-} \leqslant 100$ mg/L and $c_{CN^-} \leqslant 5$ mg/L. CN^- could be easily decomposed into N_2 and CO_x during a high temperature (>773 K) process. Therefore, no CN^- was detected in the leach solution in all the experimental cases. For F^-, it also met the environment requirement as the c_{F^-} content in the leach solution was less than 100 mg/L[96]. Therefore, by-product of the reduction-sulfurization smelting process, cleaned slag, is easy to be safely treated and directly landfilled. Although, both the leach solutions of the cleaned slag with and without CaO could meet the environmental requirement, the reasons are different. Without CaO addition, most of the fluorides volatilized out of the slag leaving a low soluble fluoride content, while when adding CaO, most of the fluorine was fixed in the cleaned slag in the form of CaF_2 which is non-volatile and insoluble. Since parts of the fluorides volatilized into the air during the smelting process and may cause pollution problems, the total fluorine including soluble and insoluble fluorine was measured to investigate the ratio of fluorine fixed in the slag.

Table 2-15 Soluble c_{F^-} and c_{CN^-} contents in cleaned slag leach solution for different SCCB additions

	SCCB addition(wt%)	4	8	12	16	20
Without CaO	F^- content in leaching solution(mg/L)	1.46	7.44	9.68	16.5	38.2
	CN^- content in leaching solution(mg/L)	ND	ND	ND	ND	ND
With 10% CaO	F^- content in leaching solution(mg/L)	1.32	6.87	9.11	15.4	35.2
	CN^- content in leaching solution(mg/L)	ND	ND	ND	ND	ND

The solution for measuring total fluorine in the cleaned slag was prepared by a mixed alkali fusion approach[97]. 0.1 g of cleaned slag or 0.06 g of SCCB, 1 g of sodium dioxide and 2 g of sodium hydroxide were mixed and rapidly heated to 923 K for 15 min and then cooled. The cooled material was leached by 50 mL of hot water(with 1 mL of absolute alcohol) and boiled for 3 min and cooled to room temperature, and then distilled water was used for filling the volume to 100 mL(V_1). Taking 10 mL(V_2) of the clarified solution into a 50 mL flask, its pH was adjusted to 5.4 with bromocresol green (indicator), nitric acid and sodium hydroxide solutions, and 15 mL of total ionic strength adjustment buffer(TISAB) and then distilled water was added to 50 mL(V_3). The mass fraction of total fluorine in the materials, F, was expressed as follows:

$$F = \frac{c_{F^-} \times V_3}{m} \cdot \frac{V_1}{V_2} \times 100\% \qquad (2\text{-}59)$$

where m is the mass of the cleaned slag, g; c_{F^-} is the concentration of fluorine ion, g/L. The fixed ratio of fluorine(%), η, is the proportion of fluorine in SCCB transferred into the cleaned slag and is expressed as

$$\eta = \frac{F_{cs} \times M_{cs}}{F_{SCCB} \times M_{SCCB}} \times 100\% \qquad (2\text{-}60)$$

where F_{cs} and F_{SCCB} are the mass fraction of total fluorine (including that from water-soluble fluorides, such as NaF and AlF_3 and water-insoluble fluorides, such as Na_3AlF_6 and CaF_2) in the cleaned slag and SCCB, respectively (%); M_{cs} and M_{SCCB} are the masses of the cleaned slag and SCCB, respectively (g).

Fig. 2-55 shows the mass fraction of total fluorine in the cleaned slag and the fixation ratio of fluorine for different SCCB additions with and without CaO. The F_{cs} increased with an increase in SCCB addition, indicating that a large part of fluorine was transferred to the slag phase during the smelting process. The F_{cs} with CaO was obviously higher than without CaO. η without CaO addition was just from 30% to 50%, indicating that about half of the fluorine was volatilized to the air. When CaO was added, more than 90% of fluorine was fixed in the cleaned slag. The rest fluorine will enter into the tail gas. Therefore, the removal of fluorine in tail gas should be considered. In industry, lime/limestone is generally used for desulphurization of flue gas[98]. At the same time of desulfurization process, fluorine will react with lime and remove from the gas[99-100].

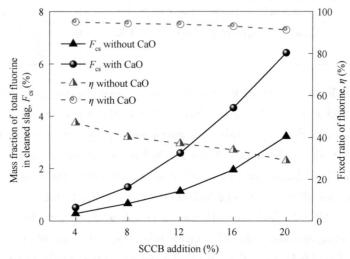

Fig. 2-55 Mass fraction of total fluorine in cleaned slag and fixation ratio of fluorine for different SCCB additions with and without CaO

Cyanide will decompose into the gases of carbon dioxide and nitrogen oxides at high temperature. The cyanide content in SPL is very low, so the gas denitration can be ignored. Therefore, there is no need to add additional process for gas defluorination.

According to the XRD pattern shown in Fig. 2-56, five main mineral phases were identified in the cleaned slag, including fayalite (Fe_2SiO_4), hercynite (Al_2FeO_4), pyroxene [$Ca(Mg, Al, Fe)Si_2O_6$], aegirine ($NaFeSi_2O_6$) and fluorite (CaF_2) when adding CaO, while no fluorine or sodium contained phase was found without CaO addition. According to the reaction (2-54) to reaction (2-56) shown in Table 2-14, the reaction of NaF, and AlF_3 with CaO can occur at the smelting temperature. Compared with the results without CaO, the content of total fluorine (including soluble fluorine and insoluble fluorine) in the slag phase increased greatly when adding CaO, which indicated that more fluorine was converted into CaF_2 and remained in slag phase. In addition, Fig. 2-54(d) showed that the recovery ratio of copper and cobalt was also increased when adding 10%(wt) CaO, which indirectly proved that the presence CaF_2 improved the slag-matte separation conditions.

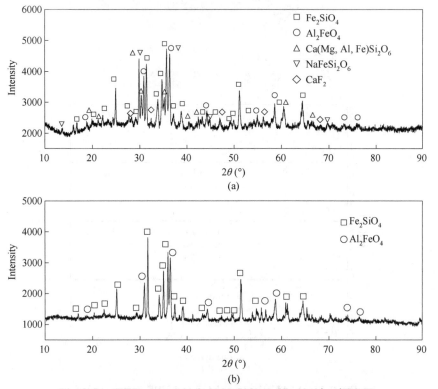

Fig. 2-56 XRD patterns of cleaned slag with 12%(wt) SCCB
(a) With 10%(wt); (b) Without CaO

From the SEM-EDS images in Fig. 2-57 and Fig. 2-58, the main phases in the cleaned slag were hercynite (area A) and fayalite (area C), and between them there was a glass phase made up of pyroxene, aegirine and fluorite (area D). Also a small amount of unseparated ferrous disulfide was found in the cleaned slag, shown at area B in Fig. 2-57 and Fig. 2-58. Almost no Cu or Co containing phases were detected in the cleaned slag with CaO addition (area D in Fig. 2-57), illustrating that copper and cobalt in the original converter slag were efficiently recovered in the Cu-Co matte/alloy phase. The Na and F in SCCB were also well fixed in the cleaned slag from NaF as aegirine and fluorite.

Characterization of metal/matte products. The smelting products are the mixture of copper matte and Co-Fe alloy. The composition and phases of the smelting product, matte/alloy, were characterized by XRD shown in Fig. 2-59. Both the matte/alloy phases

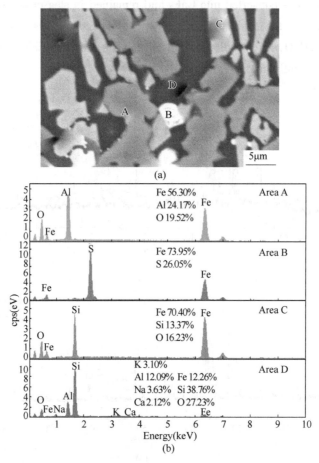

Fig. 2-57 SEM micrograph (a) and EDS pattern (b) of cleaned slag with 12% (wt) SCCB and 10% (wt) CaO

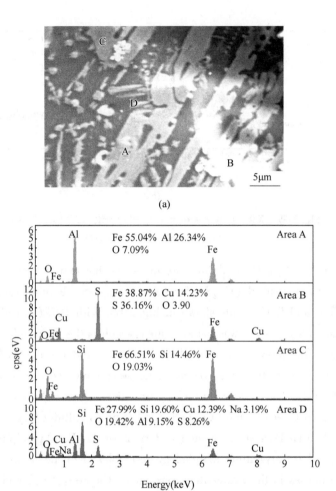

Fig. 2-58 SEM micrograph (a) and EDS pattern (b) of cleaned slag with 12%(wt) SCCB and without CaO

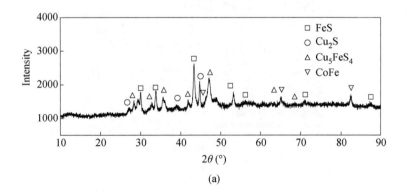

(a)

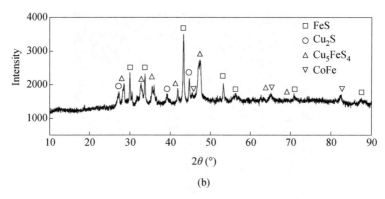

Fig. 2-59 XRD patterns of matte/alloy with 12%(wt)SCCB
(a) With 10%CaO; (b) Without CaO

obtained with and without CaO addition are the same that mainly comprised cobalt-iron alloy(Co-Fe), cuprous sulfide(Cu_2S), ferrous sulfide(FeS), bornite(Cu_5FeS_4).

From the SEM and EDS results shown in Fig. 2-60. With 10%(wt) addition of CaO, clear cobalt-iron alloy(Co-Fe) was found in area, which is the main independent mineral phase of cobalt in the product. After crushing the mixed products, the Co-Fe alloy can be easy to separate from matte phase by magnetic separation because of its magnetism. Ferrous sulfide(FeS) that collects Cu and Co from the slag was observed at area B. The main phase at area C was bornite(Cu_5FeS_4) and cuprous sulfide(Cu_2S), which are the major phases. At area D, a small amount of ferrous oxide(FeO) precipitated in bornite was observed, and its crystalline morphology resembles fishbone or dendrites. Without CaO, the main phases in the matte/alloy were mixed Cu_5FeS_4, Cu_2S and FeS, as shown at area A to D in Fig. 2-61.

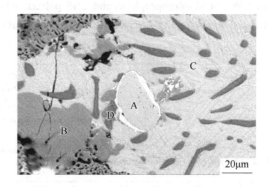

(a)

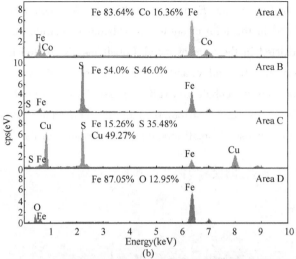

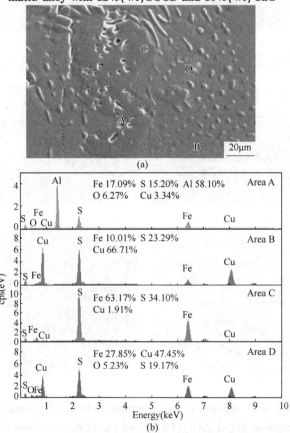

Fig. 2-60 SEM micrograph(a) and EDS pattern(b) of matte/alloy with 12%(wt)SCCB and 10%(wt)CaO

Fig. 2-61 SEM micrograph(a) and EDS pattern(b) of matte/alloy with 12%(wt)SCCB and without CaO

The distribution of Cu, Co and Fe in the smelting product could be seen in Fig. 2-62. Copper mainly existed in the form of sulfide, distributed in bornite and chalcocite; a small amount of copper existed in the form of metal. Cobalt mainly existed in the metallic state, most of which was iron cobalt alloy, and a small part of cobalt was in the sulfide form, distributed in ferrous sulfide, cobalt iron sulfide phase, bornite and chalcocite. Iron mainly existed in the sulfide phase and distributed in bornite and chalcocite, and ferrous sulfide and cobalt iron sulfide phase. A small amount of iron existed in the Fe-Co metal alloy.

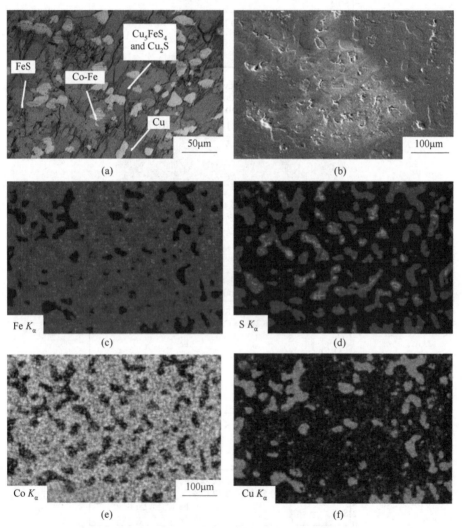

Fig. 2-62 Optical microscope image (a) and EDS mapping images of Fe(c), S(d), Co(e) and Cu(f) elements from SEM micrograph (b) of matte/alloy product

SPL can be used as an alternative or partial reductant for reduction-sulfurization smelting process of copper converter slag. However, its potential impacts need to be considered carefully. During the smelting process, sodium and fluorine in SPL may transfer into the slag in the form of sodium oxide and calcium fluoride, respectively, which may drop the slag viscosity markedly and improve the fluidity of slag. Thus, the addition of SPL should be controlled to prevent the risk of lining erosion caused by excessive Na_2O and CaF_2 content. Inevitably, a small amount of fluoride will enter into the tail gas, which needs to be concerned in the desulfurization process.

In this study, reduction-sulfurization smelting was adopted for the recovery of copper and cobalt from copper converter slags by using spent pot-lining as the reductant. The harmless reuse of spent pot-lining and the recovery of copper and cobalt from copper slags are realized, which has great economic and environmental benefits for copper and aluminum industry. The main conclusion are as follows:

(1) The thermodynamic calculations and experiments proved the feasibility of using spent pot-lining as the reductant to extract copper and cobalt from converter slags at a smelting temperature of 1673 K.

(2) Both soluble fluoride and cyanide contents in the cleaned slag could meet the environment requirement in that soluble fluoride and cyanide ions in the standard leach solution were less than 100 mg/L and not detected.

(3) A high metal recovery ratio was obtained by adding 12% (wt) spent cathode carbon block or spent silicon carbide side block. The highest recovery value was 98.7% for both copper and cobalt, when 12% (wt) spent cathode carbon block with 10% (wt) calcium oxide was added.

(4) Adding 10% (wt) calcium oxide with spent cathode carbon block could well fix the unstable fluoride in the cleaned slag in the form of calcium fluoride to avoid fluoride volatilization and dissolution, besides improving the slag property. The rest fluorine in the tail gas can be removed along with tail gas desulfurization process.

2.5.6 Recycling of spent potlining by different high temperature treatments

Spent potlining(SPL) is generated at the end of the service life of aluminum reduction cells and is a hazardous solid waste that negatively impacts the environment and ecology. Based on the differences in the hazardous and valuable components of SPL, a classified recycling and utilization path is proposed in this work. Three different high-temperature processes of oxygen-controlled treatment, vacuum treatment, and co-processing in a cement kiln were adopted for spent cathode carbon block(SCCB), spent silicon carbide block (SSCB) and spent barrier and insulating refractory materials (SBIM), respectively. After the classified treatments, all the cyanides in SCCB were decomposed, almost all the Si_3N_4 was decomposed and Na_2SiO_3 was removed from SSCB, and over 99% of the fluorides was separated from SCCB and SSCB to obtain high graphitization carbon and silicon carbide. The fluorides in SBIM were converted into a mineralizer that lowered the burning temperature of the cement clinker. The performances of prepared cement met the requirements of Portland slag, Portland pozzolana and Portland flash ash cements.

The world's primary aluminum output reached 64 Mt in 2019(International Aluminum Institute, 2020). With the increased primary aluminum output, the emissions of solid wastes have also grown explosively. The environmental and ecological issues have become one of the important bottlenecks restricting the sustainable development of alumina and aluminum industries[101-104]. The spent pot-lining(SPL) arises at the end of service life of aluminum reduction pot. For the production of every tonne of primary aluminum, about 24-30 kg of SPL is generated[105]. Thus, SPL generation reached about 1.7 Mt in 2019.

SPL usually contains high levels of soluble fluorides and cyanides. The average contents of F^- and CN^- in the leach solution of SPL are about 2000-6000 mg/L and 10-20 mg/L, respectively. It will have great ecological and environmental risks if left without reasonable treatment and utilization. At present, there are no good satisfactory technologies for SPL. Treatment in landfill and harmless disposal for SPL are the most adopted approaches but these are associated with high energy-consumption, high environmental risk and high running cost.

Many researchers focus on the treatment of the SPL by hydrometallurgical technologies, which mainly include high temperature hydrolysis, alkali leaching and calcification deposition[106], and acid leaching[107]. Lisbona et al. proposed a water-Al^{3+} two stage leaching and a nitric acid leaching approach for extracting fluorides that can be used in AlF_3 production, but the treatment and utilization of leach residue was not mentioned. Li et al. employed a three-step process for extracting and separating the fluorides, such as NaF, Na_3AlF_6 and CaF_2. The leach residue could be further treated to recover a carbon product with a purity of 95.5%. Xiao et al.[108] adopted a hydrothermal acid-leaching method to co-treat SPL and coal gangue by hydrothermal acid-leaching method and the leach residue was used for preparing silicon carbide powders. These methods are suitable for materials with high carbon content such as the first cut of SPL. Generally, the hydro-processes for SPL are lengthy and generate waste water during the processing. Moreover, flammable and explosive gases are released when SPL meets water, which makes this treatment method more dangerous. Simple alkali or acid leaching methods do not completely remove the fluorides from the SPL and do not meet the requirements for harmless treatment. The waste liquid from processing the SPL contains a large amount of acid or alkali, which requires additional sewage treatment equipment, increasing the processing cost. Further, the fluorides such as NaF and AlF_3 and slag recovered by the hydro-processing cannot be reused because they usually contain a lot of impurities. In the existing industrial operations, most of the hydro-processes do not completely achieve the harmless treatment of hazardous components and recycling of valuable components in SPL.

Many pyro-processes have also been studied. One of the pyro-processes is to treat the SPL in a rotary kiln by adding different additives (such as lignite and limestone) at a high temperature of 1000 ℃. Other researchers were focused on the utilization of carbon in SPL as a fuel. Sun et al.[109] investigated the co-combustion of SPL together with textile dyeing sludge (TDS). CaO was used for adsorbing and reducing the air pollutants from the combustions. The residue produced from these processes are landfilled or used as feedstock for building materials[110]. Givens verified the feasibility of using spent pot-lining as a fuel supplement in coal-fired utility boilers. Flores et al. evaluated the possibility of utilizing SPL as an alternative fuel in the blast furnace ironmaking process. The highly graphitized carbon in SPL can be used as a fuel, but it has a high ignition temperature. Rustad et al. and Renó et al. proposed the use of SPL in the cement industry to absorb wastes from other industrial processes[111]. Wang et al.[112] presented a vacuum distillation process (VDP) to separate the carbon from alkali metals and electrolytes in SCCB. Pyrolized SCCB with a carbon content of 91% was obtained by VDP at a 1200 ℃. There were also some industrial research on processing all the three wastes of SPL together. SPL was treated with CaO in a high temperature rotary kiln and most of the soluble fluorine was fixed in slag and the cyanides was decomposed effectively. However, about 1.14 t of residue would be generated by processing 1 t of SPL and the residue that contains carbon, fluorides and aluminosilicates could not be reused. Since the compositions of these three materials are different, they need to be detoxified according to their characteristics. Moreover, the products obtained after mixing the three materials have many impurities and cannot be reused economically.

Both the hydro-and pyro-processes without classification can be used for harmless treatment of SPL. However, from the technical and economic points of view, the classification and separation treatments are necessary for the utilization of SPL. The SPL can be divided into spent cathode carbon block (SCCB), spent silicon carbide block (SSCB), and spent barrier and insulating refractory materials (SBIM). SCCB is the contaminated carbon-rich lining material that is removed is referred to as "first cut". Both SSCB and SBIM are the contaminated refractories rich lining materials are referred to as "second cut"[113]. These solid hazardous wastes are generated at different positions that are under different environments in the reduction cell. Thus, the erosion mechanisms of SCCB, SSCB and SBIM in SPL and their contents of toxic substances, especially the fluorides, are significantly different, besides their distinct compositions. The disposal and treatment of SPL without classification cannot recycle the valuable compositions separately and are difficult to implement from the technical and economic viewpoints[114]. The key point in the SPL treatment is to develop an economic and technically sound recycling technology based on the hazardous and valuable characteristics of the SPL.

Previous investigations on the hazardous characteristics of SPL and the fluoride removal kinetics mechanism at high temperatures have been done at the University of Science and Technology Beijing. In this work, a classified high temperature treatment method was proposed to remove the hazardous components and recover the valuable components in SPL and larger scale experiments (10 kg materials) have been performed. Three different high-temperature treatment processes, i.e. the O_2-controlled process, the vacuum treatment, and the co-processing in cement kiln, were performed for SCCB, SSCB and SBIM, respectively. After the classified high-temperature processing, various valuable components such as high graphitized carbon, silicon carbide, fluoride salts and aluminum silicate were effectively recycled. The economic benefit was also discussed and compared with the conventional hydro-and pyro-processes without SPL classification. The technical and economic benefits of the classified processes proposed in this work are considerable. The purpose of this work was to removing the toxic soluble fluorides and cyanides, and to maximize the use of valuable of components carbon, SiC and aluminosilicates. The classified high temperature treatment of SPL provides feasible and economic processes for green and clean production in the aluminum industry.

The three types of solid wastes SCCB, SSCB and SBIM in SPL are exposed to different environments, and are eroded and damaged in different ways during their service lives. Due to the differences in compositions of SCCB, SSCB and SBIM as shown in Table 2-16, the SPL should be classified and treated in different ways according to the characteristics of valuable and hazardous components of each waste rather than directly without classification. The treatment and recycling of SPL should also be in technically sound and economic ways. This work proposes three reasonable treatment and utilization paths.

Table 2-16 Compositions of SCCB, SSCB and SBIM in waste (%)

Project	SCCB	SSCB	SBIM
C	55-75 Graphite	15-25 SiC	—
Na	5-15 NaF, Na_3AlF_6	0.1-8 NaF, Na_2SiO_3	0-20 Na_3AlF_6, NaF, $NaAlSiO_4$
F	5-20 NaF, Na_3AlF_6, CaF_2	0-5 NaF, CaF_2	0-30 NaF, Na_3AlF_6, CaF_2
Ca	0.5-4 CaF_2	<0.3 CaF_2	0-2 CaF_2
Al	5-15 Na_3AlF_6, Al_2O_3, Al(metal)	<0.2 Al_2O_3	20-40 Al_2O_3, Na_3AlF_6, $NaAlSiO_4$
Fe	<1 Fe_2O_3	<0.1 Fe_2O_3	0.5-2 Fe_2O_3
Si	<1 SiO_2	55-75 SiC, Si_3N_4, SiO_2, Na_2SiO_3	5-25 SiO_2, $NaAlSiO_4$
N	—	3-10 Si_3N_4	—
CN^-	0.002-0.15 NaCN	0.001-0.002 NaCN	0.0015-0.01 NaCN

Fig. 2-63 shows a schematic diagram of the classified high temperature treatment processes for SCCB, SSCB and SBIM. The main components of SCCB were carbon with high graphitization, sodium, aluminum and calcium fluorides and trace amounts of cyanide. In this study, the electric resistance heating (ERH) method was used to separate and recover a high carbon content product (CP) and fluorinated salts. During the high temperature processing, the partial pressure of oxygen was controlled to ensure the decomposition of the cyanides. However, the O_2 partial pressure should be controlled to a very low value (less than 3×10^{-2} atm) to avoid the oxidation of carbon in the SCCB and ERH lining. The SSCB was composed of silicon carbide, silicon nitride, fluoride salts and a small amount of sodium silicate. The vacuum heating method was adopted to achieve the volatilization of fluorides and the decomposition of silicon nitride and sodium silicate, and obtain pure silicon carbide. The main components of SBIM were aluminosilicates and fluorides. The best recycling method for SBIM with aluminosilicates and fluorides was co-processing in the cement clinker production system. All the aluminosilicates in SBIM were used, and the fluorides were converted into a mineralizer that lowered the burning temperature of cement.

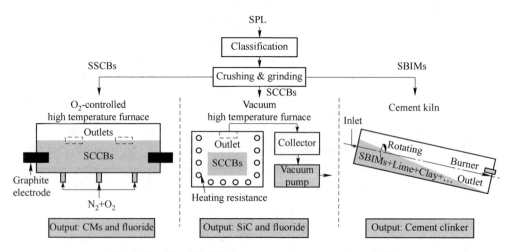

Fig. 2-63 Classified high temperature processing of SPL

An O_2-controlled high temperature treatment was performed for recycling the CP and fluorides from SCCB. The raw materials of SCCB were crushed to about 30–50 mm and placed into a resistance heating furnace[115]. Two electrodes were inserted into raw materials from both sides. Two outlets of the volatilized fluorides were set at the furnace side wall and the SCCB was evenly placed from the furnace bottom to the outlets level. An air-cooled collector was connected to the furnace outlets for recycling volatilized fluorides. The furnace lining was made of graphitized carbon to avoid corrosion by volatilized fluorides. The mixing gases of nitrogen and oxygen were injected from three bottom gas inlets. The O_2 volume fraction of mixing gas was controlled to less than 5% at the beginning of the injection. After all the cyanides are oxidized, the injection of O_2 was stopped. The injected O_2 was also mostly consumed by carbon in SCCB, and thus the graphitized carbon lining was hardly oxidized during the process. The temperature with a precision of ±1% was monitored through a sheathed thermocouple inserted into the furnace and the raw materials were heated to above 1700 ℃ for 5 h. After O_2-controlled high temperature treatment, the chemical and phase compositions and microstructures of output materials of CP and fluorides were analyzed.

A high temperature vacuum treatment was performed for recycling the SiC and fluorides from SSCB. The raw materials of SSCB were crushed and ground to smaller than 5 mm and placed into a vacuum heating furnace. The furnace lining was also made of graphitized carbon. An outlet of the volatilized fluorides was set up at the top of the furnace. The SSCB material was heated by molybdenum heating elements and the temperature was controlled to (1700±1) ℃ for 2 h. The pressure was auto-controlled to lower than 10^{-3} atm. After high temperature vacuum treatment, the chemical and phase composition and microstructure of output materials of SiC and fluorides were analyzed.

1% (wt) of the SBIMs was added as the raw materials to prepare the cement clinker. All the raw materials were crushed and ground to smaller than 80 μm and made to test cake with water. Then, the prepared test cakes were placed into muffle furnace headed by a silicon molybdenum heating resistance and heated to 1250–1450 ℃ for 2 h to obtain the cement clinker. The calcined cake (cement clinker) prepared at 1250 ℃ was ground and mixed with 5% (wt) gypsum to prepare Portland cement. The prepared pellets or test cakes were kept at 1450 ℃ or 1500 ℃ for 0.5–2 h, and then discharged from the furnace. The free-CaO content in the cement clinker prepared at different temperature and the cement performance were analyzed.

A standard leaching solution for measuring solubilized fluoride and cyanide contents was prepared by stirring 100 g solid materials in 1 L of water for 1 h. Total solubilized fluorine in the SPL and the output materials was determined potentiometrically using a fluoride ion-selective electrode (ISE, PXSJ-216, Shanghai INESA Scientific Instrument, China) that had a minimum detection concentration of 0.05 mg/L. The cyanide content was measured by a silver nitrate titration method (SEPA), which provided a minimum detection concentration of 0.025 mg/L. The phase composition was determined by X-ray diffraction analysis using a PW 1710 diffractometer (XRD, Philips, The Netherlands) and the microstructure was observed using a JSM-7500 F scanning electron microscope

(SEM, JOEL, Japan). The standard Gibbs free energies and boiling points were calculated by Outotec HSC Chemistry 9 software. The cement performances of normal consistency, soundness, setting time and rupture and compressive strengths were measured according to IS 4031:1988[116].

Cathode carbon blocks (CCB) are installed as the first layer of lining at the bottom of the reduction cell. The CCB are scoured and eroded by the molten aluminum and electrolyte during the reduction process. Besides, the CCB deform and crack due to the long-term influence of thermal stresses. During the service life of CCB, a large amount of the fluoride salts penetrates into the cracks, with NaF permeating especially deeply. As shown in Fig. 2-64(a), another toxic component in the SCCB is the cyanides that formed near the cathode collector bars where Na permeates into carbon and reacts with N_2 in the air. Some Al_4C_3 may be generated by the reaction among the cathode carbon, molten aluminum and electrolyte. The graphitization degree of the CCB may increase to above 90% during the service life. Therefore, the key point for the recycling of SCCB was to achieve the decomposition of cyanides and the separation of fluorides and CP.

The fluorides can volatilize and separate from the CP during the high temperature processing. According to the TG/DSC results, the volatilization of fluorides occurred only when the temperature was above their respective melting points, which was a steady and slow process under both Ar and Ar-O_2 atmospheres. As shown in Fig. 2-64(b), most boiling points of fluorides, except CaF_2, were below 1700 ℃.

According to the previous work, the main phases of fluorides made up of NaF, Na_3AlF_6 and a small amount of CaF_2 did not change during high temperature processing. The contents of LiF and KF were very low, and their phases were not detected in XRD patterns. Because of the different boiling points, the fluorides volatilized successively during the heating process. AlF_3 present at a low content volatilized first at 1283 ℃. The volatilization of NaF and decomposition of Na_3AlF_6 occurred at 1660-1690 ℃. There might also be a small amount of CaF_2 particles mixed with the volatilized gas in the cooled collector. As the temperature cooled while the fluorides traveled to the collector, the reverse reaction of Na_3AlF_6 formation occurred between NaF and AlF_3.

Over 70% of the SCCB was highly graphitized carbon that had good conductivity and easily self-heated when electrical current was applied. The NaCN in the SCCB was oxidized by adding oxygen in the high temperature treatment process, as show in Fig. 2-64(c). The types of the decomposition products depended on the oxygen partial pressure. Thus, the CN^- was oxidized to CO_2 and N_2 at a low O_2-pressure. According to the previous research on TG/DSC-MS analysis using an ECSA® method[117] in an Ar-O_2

atmosphere, more than 75% of the cyanides was converted to N_2 and to the rest was converted to NO. No N_2 was generated under an Ar atmosphere. In order to avoid the generation of NO_x, the oxygen partial pressure in the furnace should be controlled at a low level. Thus, the CN^- could be converted to CO_2 and N_2 at a low O_2 pressure according to the reaction of $4NaCN + 5O_2 = 2Na_2O + 4CO_2 + 2N_2$. The flow sheet of the oxygen-controlled high temperature treatment process of SCCB is shown in Fig. 2-64(d).

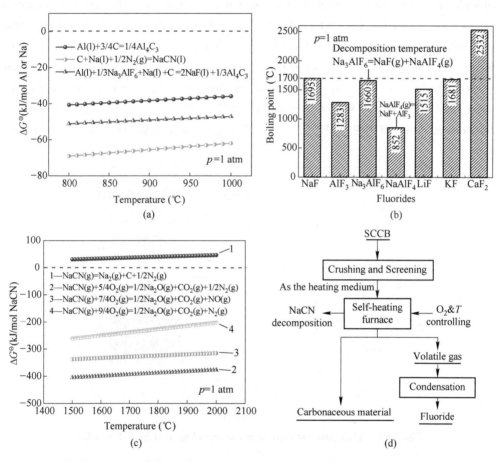

Fig. 2-64 Thermodynamic calculation and flow sheet of oxygen-controlled high temperature treatment for SCCB

(a) Standard Gibbs free energies of NaCN and NaF generation reactions; (b) Boiling points of different fluorides;
(c) Standard Gibbs free energies for decomposition reactions of NaCN; (d) Processing flow sheet

Fig. 2-65 and Table 2-17 show the changes of phases and compositions between the SCCB and CP after 5 h of O_2-controlled high temperature treatment. During the high temperature processing at ≥1700 ℃, the fluorides separated from the SCCB and the fluorine contents were decreased from 7.2% in SCCB to 0.00915% in CP. The fluorine content in the leach solution was <10 mg/L, which was lower than the limit of 100 mg/L in the identification standards for hazardous wastes. The NaCN was decomposed completely and no cyanide was detected in the CP. From the SEM micrograph shown in Fig. 2-65, there were some pale crystals which were fluorides distributed in the SCCB. After treatment, the surface of the carbon product is cleaner, which indicated that the fluorides were separated effectively at the processing temperature. The recovered carbon product was with a low ash content of 1.6% and a high graphitization degree of 96% and its fixed carbon content reached 98.4%. The residual CaF_2 exists as an ash component in the CP. The CP is a good raw material for preparing graphitiferous cathode carbon block, cathodic pastes, and the carbon block for furnace lining for which the ash content is required to be ≤2%-8%, ≤3%-7%, and ≤5%-16%, respectively. The crystallized fluorides made up of NaF, CaF_2 and Na_3AlF_6 can be reused in the aluminum smelting system.

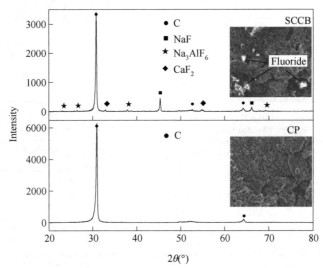

Fig. 2-65   XRD patterns and micro-morphologies of SCCB and CP

Table 2-17   Compositions of SCCB and CP   (wt%)

Project	SCCB	CP
Fixed carbon	70.3	98.4
Ash	27.9	1.6
Soluble fluorine	7.12	0.00915
Cyanide	16 mg/L	ND
Graphitization	91.1	94.8

The silicon carbide blocks (SCB) with high corrosion resistance and low electric conductivity are installed as the side lining of the reduction cell. During the service life, the SiC and Si_3N_4 may be oxidized and react with NaF and Na_3AlF_6 to form Na_2SiO_3 and SiF_4 gas. The key point in the recycling of SSCB was to achieve the removal of Na_2SiO_3 and the decomposition of Si_3N_4 to obtain a usable silicon carbide product. The cyanides are not generated near the side lining, and thus there is almost no cyanides in the SSCB.

The cyanide content in SSCB is very low, less than 0.002%. The main concern of the toxic component in SSCB is fluorides. The cyanide can be easily oxidized and decomposed at a very low oxygen partial pressure during the vacuum heating and cooling processing, i. e. no additional O_2 is needed. In addition, the SSCB with poor conductivity cannot be self-heated as an electrode like SCCB. Thus, it is difficult to heat SCCB to a high temperature above 1700 ℃ with external heating. At relatively low temperatures, lowering the pressure can decrease the boiling points of fluorides, and thus vacuum heating was adopted to improve fluoride volatilization.

As shown in Fig. 2-66(b), the boiling points of fluorides, except CaF_2, can be lowered to <1000-1200 ℃ under a vacuum of $10^{-4}-10^{-3}$ atm. The decomposition of Si_3N_4 and the reduction of Na_2SiO_3 can be carried out at ⩽1515 ℃ and $p ⩽ 10^{-3}$ atm. Thus, a vacuum furnace was suggested for the processing of SCCB and the flow sheet is shown in Fig. 2-66(d). After high temperature vacuum processing of SSCB, silicon carbide and fluorides including NaF, Na_3AlF_6 and CaF_2 that can be reused for preparing refractories and electrolytes, respectively, were obtained.

Fig. 2-67 and Table 2-18 show the changes in phases and compositions between the SSCB and recycled SiC after an hour of high temperature vacuum treatment. During 1600 ℃ high temperature processing, the fluorides completely volatilized and separated from the SSCB. The vapors were pumped from the high temperature zone to a condenser. The inner lining of the volatilizer also needed to be made of carbon because of the corrosiveness of fluoride gases. After the high temperature vacuum processing, the fluorine content in SiC decreased from 2.3% in SSCB to 0.0092%. The soluble fluorine content in the leach solution was <10 mg/L, and there was no SiN_4 detected in the SiC materials. As shown in Fig. 2-67, the main phase in the product was SiC, and a small amount of Si was observed and microscopic morphology showed that the surface of the SiC is very clean without silicon nitride and fluoride impurities after a vacuum treatment at 1700 ℃. The purity of SiC was up to 94.4%, which makes it suitable for reuse in the for production of silicon carbide refractories.

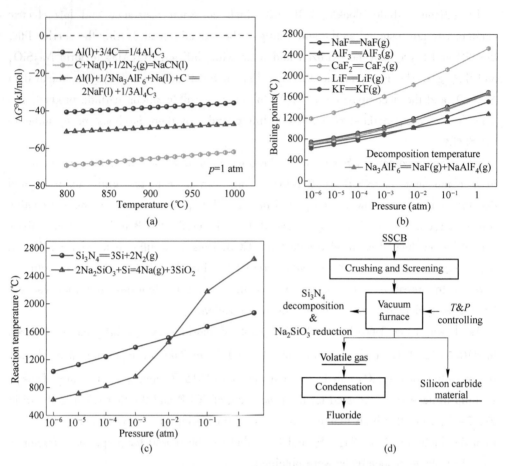

Fig. 2-66 Thermodynamic calculation and flow sheet of
high temperature vacuum treatment for SSCB

(a) Standard Gibbs free energies of the reactions of SSCB; (b) Boiling points of fluorides;
(c) Decomposition temperatures of Si_3N_4 and Na_2SiO_3 under different vacuum levels;
(d) Processing flow sheet

Table 2-18 Compositions of SSCB and recycled SiC   (%)

Project	SSCB	SiC
SiC	76.7	94.4
Fluorine	2.3	0.0092
Na_2SiO_3	3.1	0.90
SiO_2	4.2	0.47

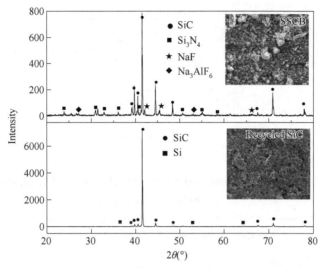

Fig. 2-67 XRD patterns and micro-morphologies of SSCB and recycled SiC

Compared with SCCB and SSCB, the value of SBIM is lower, and their most economic and reasonable use is to add to the cement production system as raw materials. As shown in Fig. 2-68(a), the main components of the SBIM are the same as those of the cement clinker. During the cement sintering process, the fluorides react with lime and turn into fluorite, which is a good mineralizer for cement sintering. The main fluoride compounds in SBIM were NaF and Na_3AlF_6. According to the thermodynamic calculation shown in Fig. 2-68(b), CaF_2 may also be generated during the clinker sintering process. These fluorides are good mineralizers that can accelerate the reaction process, decrease the clinker sintering temperature, and improve the cement[118]. The flow sheet for co-processing SBIM in the cement kiln is shown in Fig. 2-68(c).

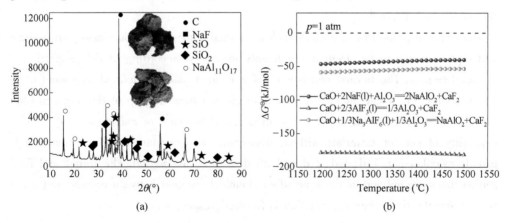

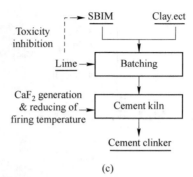

(c)

Fig. 2-68 Thermodynamic calculation and flow sheet of co-processing of SBIM in cement kiln

(a) XRD pattern of SBIM; (b) Standard Gibbs free energies of the reactions between CaO and fluorides; (c) Processing flow sheet

The cement clinker was prepared by heating the mixed materials to 1250–1450 ℃ for 2 h. The calcined cake (cement clinker) prepared at 1250 ℃ is ground and mixed with 5% (wt) gypsum to prepare Portland cement. The free CaO content is one of the most important factors to measure the chemical properties of cement clinker. The free CaO content was determined by a glycerine-ethanol method with an error of less than 0.1%. 0.5 g sample was mixed with 15 mL of anhydrous glycerine-ethanol and heated to boiling. Phenolphthalein was used as the indicator and 0.1 mol/L toluene acid absolute ethanol solution was used for titrating sample solution until to be neutral. The free CaO content was calculated by

$$f_{CaO} = \frac{c_{CaO} V}{m \times 100} \times 100\%$$

where c_{CaO} is the mass of the CaO per milliliter toluene acid absolute ethanol solution, mg/L; V is the consumed volume of the toluene acid absolute ethanol solution, mL; m is the mass of solid sample, g.

The physical properties of clinker such as normal consistency, soundness, setting time and rupture/compressive strength are mainly expressed according to the properties of Portland cement. The normal consistency of a cement paste is defined as water (wt), which permits the Vicat plunger to penetrate 10 mm from the top of the Vicat-mould. Flexural or compressive strength is the mechanical measure of maximum load bearing capability of cement materials without undergoing any permanent deformation and is usually calculated in 1 d, 3 d, 7 d and 28 d after curing period. The initial and final setting times are defined as the time when chemical reaction between cement and water is start and the time when the reaction is finished, respectively.

Fig. 2-69 shows the content of free calcium oxide in the burned cement clinker. When 1%(wt) SBIM was added, the free CaO content increased slightly as the temperature increased from 1250 ℃ to 1350 ℃. That may be because at a lower temperature, the addition of SBIM changed the sintering characteristics of the materials, that is, decreased the sintering temperature. Part of the raw materials may be sintered incompletely and formed a small amount free CaO, causing the small increase in f_{CaO} (%) at lower temperature. With a further increase in temperature from 1350 ℃ to 1400 ℃, the raw materials were calcined more effectively, the influence of added SBIM was weakened, and the free CaO content decreased. The sintering temperature of the cement clinker decreased by 100-120 ℃ after adding 1%(wt) waste refractory, which achieved a low calcium oxide content of less than 0.3%. Table 2-19 shows the properties of the cement clinker fired after adding the SBIM. The addition of 1% SBIM did not affect the quality of cement clinker, and its performance indexes of normal consistency, soundness, setting time and rupture/compressive strength met the requirements of Portland slag(PS), Portland pozzolana(PP) and Portland flash ash(PP) cements.

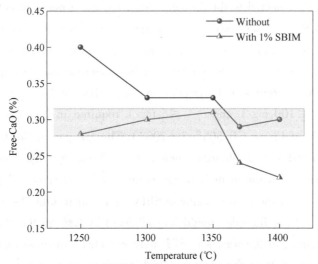

Fig. 2-69 Free-CaO contents in the cementclinker with and without SBIM(wt)

Table 2-19 Performance comparison of the cements with and without SBIM

Cement type		PS,PP,PF cement	SBIM-cement
Normal consistency(%)		24-26	23.4
Soundness		—	Qualified
Setting time(h)	Initial	1-3	2.6
	Final	5-8	6.0

Continued Table 2-19

Cement type		PS, PP, PF cement	SBIM-cement
Flexural(F) and Compressive(C) strengths(MPa)	3 d, F	2.5-4.5	3.4
	28 d, F	5.5-7.0	5.7
	3 d, C	10.0-23.0	12.5
	28 d, C	32.5-52.5	33.0

Unlike the processes without classification, each process proposed in this work produced no waste residue and no waste water. The main products carbon, SiC and fluorides are the raw materials for producing electrodes, refractories and electrolytes that can be readily reused in the aluminum industry. SBIM, the most difficult part to treat in SPL, can be used as an additive to produce cement, which completely solves the problem of waste residue storage.

As far as the toxic components in SPL, almost all the fluorine was collected during the O_2-controlled and high temperature vacuum processing. The fluorides were condensed in a cooled collector connected to the furnace outlet. The main phase of collected fluorides were NaF, Na_3AlF_6 and CaF_2 which can be reused as electrolyte products. More than 99.5% of the soluble fluorine can be removed from SCCB and SSCB, and more than 95% of the fluorine gas can be collected in the experimental cooling collector. The contents of soluble fluorides and cyanides in the carbon and SiC products meet the requirements ($c_{F^-} \leqslant 100$ mg/L and $c_{CN^-} \leqslant 5$ mg/L required in National Standard of the People's Republic of China GB 5085.3—2007). The tail gas of the high temperature processing contained a total fluorine content of 2.5 mg/m^3 that meet the emission requirement (6 mg/m^3 required in National Standard of the People's Republic of China GB 9078—1996). For the co-processing of SBIM in a cement kiln, the addition of SBIM is only 1%(wt) and the fluorides would be well fixed by CaO in the clinker during the processing. Therefore, all the wastes in SPL were recycled and reused and no secondary wastes were generated from the classified high temperature processes, which meet the requirement of green and clean production.

The comparison of economic benefits among different processes is listed in Table 2-20. Because of the toxicity of SPL, the aluminum industry cannot treat SPL by landfill or direct incineration. The processing by a qualified company may cost 300-400 U.S. dollars per tonne SPL. Some aluminum companies have treated the SPL using hydro-or pyro-processes, which may cost about 100-200 U.S. dollars per tonne SPL. Furthermore, these processes only produce high ash content carbon and low purity fluorides or waste residue that cannot be reused. In this work, SCCB, SSCB and SBIM in SPL were treated separately to produce the products that can be recycled. The thermal efficiency of the treatment furnace for SCCB is about 90% and the energy consumption is about 2 kW/h per kg SCCB. For SSCB, the fluorine content is much lower than it in SCCB, and thus the processing time for SSCB is shorter. It is estimated that the processing time of SSCB is only 50%–60% of that of SCCB. The thermal efficiency of the vacuum furnace for SSCB is about 70% and about additional 10%–20% energy consumption is required to maintain the vacuum. Therefore, the energy consumption, as well as the processing cost, of the two materials is almost the same. The net profit of this process was more than 150 U. S. dollars per tonne of SPL, which provides a much better economic benefit.

Table 2-20 Comparison of economic benefits

Project		Processing cost of SPL ($/t)	Products and revenue ($/t)	Net profits of SPL ($/t)
Landfill incineration by qualified company		300-400	no revenue	Nagative 300-400
hydro-process		100-200	high ash content carbon product and low purity fluorides, no revenue	Nagative 100-200
pyro-processes		100-200	waste residue, no revenue	Nagative 100-200
Classified high temperature treatments	SCCB 55%	150	Yield 90%; 70% high graphitized CP: 500 20% fluorides: 300	Positive 162.5[①]
	SSCB 10%	150	Yield 90%; 80% SiC: 700 10% fluorides: 300	
	SBIM 35%	70	Co-processing by cement plant, no revenue	

①Calculated by 55%×(-150+500×70%+300×20%)+10%×(-150+700×80%+300×10%)+35%×(-70+0)= 162.5.

Classified high temperature processes for different parts of spent pot-lining were proposed to achieve the harmless treatment of hazardous components and efficient recycling of valuable components.

(1) The oxygen-controlled high temperature treatment was adopted for the spent cathode carbon block. At a temperature of ≥1700 °C for 2 h, over 99.9% of the fluorides was seperated from the carbon product; Under a controlled O_2 content; the cyanide was completely decomposed; a recycled carbon product with a purity of 98.4%, a graphitization degree of 94.8%, and a soluble fluorine content of <10 mg/L was obtained.

(2) The high temperature vacuum treatment was adopted for spent silicon carbide block. At 1700 ℃ and 10^{-3} atm for 2 h, almost all of the Si_3N_4 and Na_2SiO_3 was decomposed; Over 99.9% of the fluorides was seperated from the SiC waste; andrecycled SiC with a purity of 94.4% (wt) and a soluble fluorine content of <10 mg/L were obtained.

(3) Co-processing of the spent barrier and insulating refractory materials in a cement kiln was performed to prepare burned cement clinkers. With 1% (wt) addition of spent refractory, the sintering temperature for clinker was lowered by 100-200 ℃. The performances of prepared cement met the requirements of Portland slag, Portland pozzolana and Portland flash ash cements.

2.6 Hydrometallurgy disposal technology

2.6.1 Leaching of waste cathode carbon block

Waste cathode carbon blocks are hazardous solid waste generated during the overhaul of aluminum electrolysis cells. Due to long-term erosion by electrolytes, they contain a large amount of soluble fluoride, which can cause serious environmental pollution when stored or buried. The leaching kinetics of sodium fluoride from waste cathode carbon blocks were mainly studied, revealing the effects of temperature, particle size, liquid-solid ratio, and other factors on the leaching of sodium fluoride from waste cathode carbon blocks in aluminum electrolysis. The results show that under the conditions of a liquid-solid ratio of 25 mL/g, temperature of 85 ℃, and particle size of 0.058-0.075 mm for 1 hour, the leaching rate of soluble fluorine can reach 98.9%, and the soluble fluorine content in the leaching residue is 83.53 mg/L, which is lower than the safe discharge standard of 100 mg/L. This can achieve harmless treatment of waste cathode carbon blocks. The leaching process conforms to the shrinkage core model controlled by diffusion within the solid film layer, with apparent activation energy of 8.97 kJ/mol.

Aluminum is the highest production and most widely used non-ferrous metal. With the rapid growth of raw aluminum production and the development of the aluminum electrolysis industry, the discharge of maintenance slag from the reduction cell has also been accelerated. According to statistics, for every 1 ton of raw aluminum produced, approximately 20-30 kg of overhaul slag is generated. The overhaul slag of the electrolytic cell consists of three parts: waste cathode carbon blocks, waste silicon carbide side blocks, and waste refractory materials, with waste cathode carbon blocks as the main waste. The main components of waste cathode carbon blocks include valuable components such as carbon, cryolite, and fluoride, and are classified as hazardous solid waste due to their high content of soluble fluorine and small amounts of aerosols. If discarded cathode carbon blocks are simply stored in the open air without strict protection, the soluble fluorine in them will be dissolved into the soil and groundwater during cloudy and rainy days. This process is accompanied by the release of harmful gases, which can cause environmental problems. At present, the waste cathode carbon blocks from aluminum electrolysis have been listed in the National Hazardous Waste List by the state. According to the national standard, if the concentration of soluble fluorine in the leaching solution exceeds 100 mg/L, it is considered hazardous waste. The actual soluble fluorine content in China averages 2000 to 4000 mg/L, with some reaching 6000 to 8000 mg/L, which is far higher than the discharge requirements for solid waste. Therefore, the clean treatment and resource utilization of waste cathode carbon blocks in aluminum electrolysis is an urgent problem to be solved.

At present, the treatment methods for waste cathode carbon blocks are mainly divided into fire treatment and wet treatment. The fire treatment process mainly includes Reynolds process in the UK, plasma treatment process, AUSMELT process in Australia, Comalco process, SPLIT process in France, etc. The basic principle of the Chalco SPL process proposed by China National Aluminum Corporation is to mix spent potlining with limestone or other additives and add them to a high-temperature furnace for roasting, so that soluble fluoride can generate insoluble calcium fluoride. However, in actual operation, there may be the generation of HF gas, high toxicity of soluble fluoride leaching, how to commercialize the discharged slag, and unstable process operation A series of problems, such as the combustion of high value carbon materials and the failure to effectively recycle and utilize them, have led to high operating costs for enterprises, failure to achieve the expected green, economic, and efficient treatment goals, and difficulty in large-scale promotion.

The wet treatment processes mainly include Alcoa's LCL-L process, flotation method, chemical leaching method, etc. The basic principle of the LCL-L process is to first extract soluble fluoride with water, and then extract the remaining fluoride with low caustic soda. This method has a good processing effect, but its disadvantages include a long process, a large number of processing equipment, and is not suitable for promotion and utilization in small and medium-sized enterprises. The flotation method for treating waste cathode carbon blocks in aluminum electrolysis cells has the advantages of simple operation, complete separation of carbon and electrolyte, cleanliness and environmental protection, so it has been increasingly valued in recent years. Adding alkaline leaching process before flotation can further improve the quality of the obtained product. However, the drawbacks of current flotation methods are low product purity and complex flotation conditions. The chemical leaching method mainly involves treating waste cathode carbon blocks with acid, alkali, or deionized aqueous solutions to leach the carbon and fluoride salts from them. The chemical leaching method has the advantages of low energy consumption, short process, simple operation, and low cost, but its disadvantages are the large amount of acid and alkali used and the complex leaching solution. However, there is limited research on aqueous solutions as a single leaching agent.

In response to the poor performance of traditional wet treatment, this article proposes using aqueous solution as a single leaching agent to treat aluminum electrolysis waste cathode carbon blocks. The leaching behavior and kinetics of soluble fluorine are studied, and factors such as reaction temperature, particle size, and liquid-solid ratio are explored to determine the optimal leaching conditions for soluble fluorine, so as to meet the discharge standards for non hazardous solid waste. It also provides some reference for the wet treatment of waste cathode carbon blocks in aluminum smelting enterprises.

The waste cathode carbon blocks used in the experiment were from Jiaozuo Wanfang Aluminum Industry Co., Ltd. After being crushed and mixed evenly, chemical composition analysis was conducted. In addition to containing a large amount of C, the waste cathode carbon block also contains a large amount of elements such as F, Al, Na, Ca, etc. due to the infiltration and erosion of a large amount of electrolytes during the production process. The XRD spectrum of the waste cathode carbon block is shown in Fig. 2-70. From Fig. 2-70, it can be seen that the main phases in the waste cathode carbon block are C, NaF, Na_3AlF_6, and CaF_2. The content of NaF is relatively high, which is generally generated by the reaction between infiltrated electrolytes or metallic sodium and cryolite. In areas with lower edge temperatures in aluminum electrolysis cells, the content of NaF will be higher. The average concentration of soluble fluorine in the waste cathode carbon block measured by the fluoride ion electrode method is 7500 mg/L.

The experimental instruments mainly include HX203T electronic balance, SHJ-A6 magnetic stirring water bath, PXSJ-216 fluoride ion meter, 202-1 electric constant temperature drying oven, ZBSX-92A standard vibrating screen, beaker, glass cup, measuring cylinder, etc.

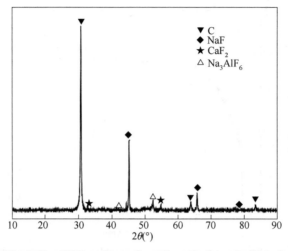

Fig. 2-70　XRD pattern of waste cathode carbon block

This experiment uses aqueous solution as the leaching agent. From Fig. 2-70, it can be seen that fluorine in the raw material mainly exists in the form of NaF and CaF_2, which are insoluble compounds. NaF is highly soluble in water, and its solubility in water is 3.66–5.08 g/L (15–100 ℃). Therefore, the soluble fluorine in the waste cathode carbon block mainly comes from the large amount of dissolution of NaF. Although NaF also undergoes hydrolysis reactions, the reaction amount is very small, and the impact can be ignored. Therefore, in theory, it is believed that the leaching rate of soluble fluorine is entirely due to fluoride ions in NaF. This experiment adopts a single factor test method to mainly study the effects of temperature, particle size, liquid-solid ratio, etc. on the leaching rate of soluble fluorine in waste cathode carbon blocks. First, place a 1000 mL flat bottomed flask in a constant temperature heated water bath, add a certain amount of water, set the temperature, and after reaching the set temperature, add 10 grams of raw materials to the flask. During the leaching reaction process, strictly control the temperature and stir with strong magnetic force. Starting from the leaching reaction, time the leaching solution with a pipette and measure the soluble fluorine content in the leaching solution using the fluoride ion selective electrode method.

Take 10 g of raw material and investigate the effect of temperature (30 ℃, 50 ℃, 70 ℃, 90 ℃) on the fluorine leaching rate under the conditions of a liquid to solid ratio of 5 (volume to mass ratio, mL/g, the same below) and a particle size of 0.106–0.15 mm. The results are shown in Fig. 2-71. As shown in Fig. 2-71, the effect of temperature on the leaching rate of soluble fluoride is still quite significant. Temperature is the main control parameter for molecular thermal motion. As the temperature increases, molecules

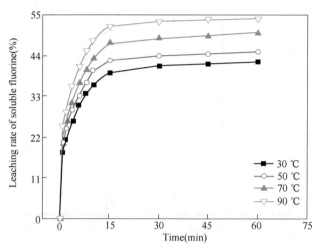

Fig. 2-71 Effects of temperature on leaching rate of soluble fluorine

gain more energy, thermal motion intensifies, the number of active molecules increases, and the number of molecules participating in the leaching reaction per unit time increases, resulting in an accelerated leaching rate. As the temperature increases, the leaching rate of soluble fluorine in the waste cathode gradually increases. When the temperature rises from 30 ℃ to 90 ℃, after 15 min of leaching, the leaching rate of soluble fluorine increases from 37.8% to nearly 52%. However, the change in the leaching rate of soluble fluorine is not particularly significant with the increase of temperature.

Take 10 g of raw material and investigate the effect of different particle sizes(0.150–0.300 mm, 0.106–0.150 mm, 0.075–0.106 mm, 0.058–0.075 mm) on the leaching rate of soluble fluoride under the condition of a liquid-solid ratio of 5 and a temperature of 90 ℃. The results are shown in Fig. 2-72. As shown in Fig. 2-72, as the particle size of the raw material decreases, the leaching rate of soluble fluorine in the waste cathode gradually increases. When the particle size decreases from 0.150–0.300 mm to 0.058–0.075 mm, after 15 min of leaching, the leaching rate of soluble fluorine increases from 39.6% to 60.1%, and the leaching rate increases significantly. This is because the smaller the solid particle size, the larger the specific surface area of the particles, the more reactions occur with the leaching solution per unit time, and the faster the leaching rate.

Take 10 g of raw material and investigate the effect of different liquid-solid ratios(5, 10, 15, 20) on the leaching rate of soluble fluoride under conditions of temperature 90 C and particle size 0.058–0.075 mm. The results are shown in Fig. 2-73. From Fig. 2-73,

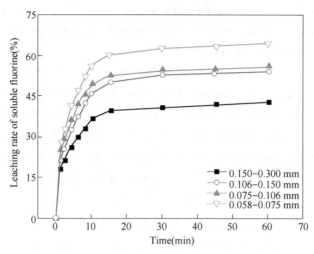

Fig. 2-72 Effects of particle size on leaching rate of soluble fluorine

it can be seen that the leaching rate of soluble fluorine in waste cathode rapidly increases with the gradual increase of liquid-solid ratio, with the liquid-solid ratio increasing from 5 to 20. After 15 min of leaching, the leaching rate of soluble fluorine increases from 61.5% to 82.4%. The leaching curve in Fig. 2-73 can be clearly divided into two stages: the first stage is that the leaching time is within 15 min. Due to the high solubility of NaF in aqueous solution, soluble fluorine rapidly dissolves in this stage, so the leaching rate of soluble fluorine increases rapidly with time in the early stage of leaching; The second stage is the later stage of leaching, with a leaching time greater than 15 min. Due to the fact that most of the exposed soluble fluorine has been leached,

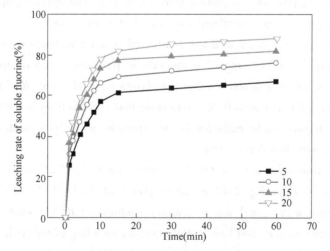

Fig. 2-73 Effects of L/S on leaching rate of soluble fluorine

it is difficult for the fluorine in the particles to be leached again. The extension of reaction time has a small impact on the leaching rate, but with the extension of time, the leaching rate still increases slightly.

The leaching reaction in this experiment mainly involves the dissolution reaction of a large amount of NaF, only involving liquid-solid phase reactions. According to the kinetics of wet gold treatment, the leaching rate control is mainly controlled by the external diffusion layer of the leaching agent solution through the surface of the ore particles, the internal diffusion control of the leaching agent through the solid film, and the mixed diffusion control. The rate equations of the control kernel model for each stage are as follows:

$$1 - (1-a)^{1/3} = k_r \times t \tag{2-61}$$

$$1 - 2a/3 - (1-a)^{2/3} = k_d \times t \tag{2-62}$$

$$[\ln(1-a)]/3 - 1 + (1-a)^{-1/3} = k_m \times t \tag{2-63}$$

where a is the leaching mass fraction of soluble fluorine, %; t is the leaching time, min; k_r, k_d, and k_m are the apparent rate constants controlled by chemical reactions, solid product layer diffusion, and their mixing, respectively.

The kinetic control model for analyzing the leaching process of soluble fluorine is mainly analyzed from two aspects: Kinetic equation and apparent activation energy. According to the graph of the effect of temperature on the leaching rate, it was found that after 15 min of leaching, the leaching rate gradually slowed down. Therefore, the leaching rate under different thermodynamic temperature conditions within the first 15 min of leaching was brought into each diffusion reaction control equation to obtain the kinetic constant k. Then, using the Arrhenius formula, the Arrhenius curve of soluble fluorine during the leaching process was plotted with $\ln k$ to $1/T$. The apparent activation energy of the leaching process was calculated from the Arrhenius curve.

Linear fitting was performed on a and t obtained at different temperatures according to Eq. (2-61) – Eq. (2-63) to determine the kinetic parameters and control steps of the leaching process. The results showed that the correlation coefficients R of Eq. (2-61) and Eq. (2-62) were both less than 0.91, indicating that the leaching of soluble fluorine in aluminum electrolysis waste cathodes is not suitable to be represented by chemical reactions and mixed diffusion control.

Plot the experimental results of Fig. 2-71 according to $[1-2a/3-(1-a)^{2/3}]$ for t, as shown in Fig. 2-74. From Fig. 2-74, it can be seen that $[1-2a/3-(1-a)^{2/3}]$ has a good linear relationship with t, and the correlation coefficient (R^2) at each temperature is above 0.955, indicating a good linear relationship. According to the Arrhenius equation:

$$k = A \times \exp[-E_a/(RT)] \tag{2-64}$$

Taking the logarithm of both sides yields the following formula

$$\ln k = \ln A - E_a/(RT) \qquad (2\text{-}65)$$

where k represents the reaction rate at different temperatures, and A represents the frequency factor; E_a is the apparent activation energy, kJ/mol; T is the thermodynamic temperature, K; $R = 8.314$ J/(mol·K).

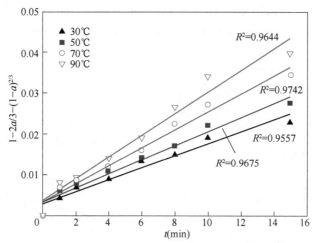

Fig. 2-74 Relationship between $[1-2a/3-(1-a)^{2/3}]$ and t at different temperatures

The $\ln k$-$1/T$ fitting curve is performed, and the results are shown in Fig. 2-75. From Fig. 2-75, the apparent activation energy E of the leaching reaction can be determined to be 8.97 kJ/mol. The apparent activation energy value further proves that the leaching process mainly conforms to the internal diffusion control model.

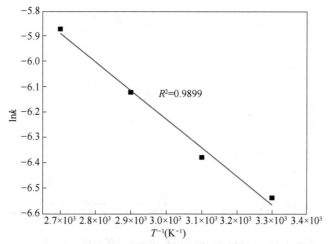

Fig. 2-75 Curve of $\ln k$-$1/T$

Plot the leaching rate at different particle sizes according to the relationship between $[1-2a/3-(1-a)^{2/3}]$ and t, as shown in Fig. 2-76. As shown in Fig. 2-76, the correlation coefficients of the fitted straight lines are all greater than 0.95, indicating a good linear relationship. The reaction rate constant k_d at each particle size is obtained; Combining the Arrhenius equation, plot $\ln d_0$ with $\ln k_d$, as shown in Fig. 2-77. The reaction order of the spent cathode particle size on the leaching reaction is -0.7615, indicating that reducing the particle size can better promote the leaching reaction.

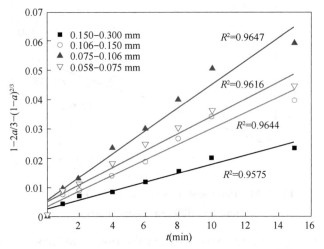

Fig. 2-76 Relationship between $[1-2a/3-(1-a)^{2/3}]$ and t under different particle sizes

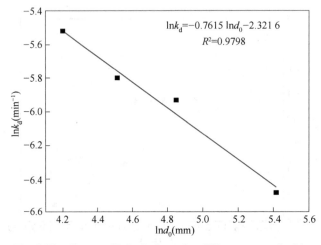

Fig. 2-77 Curve of $\ln k_d$-$\ln d_0$ under different granularities

Plot the leaching rate of different liquid-solid ratios according to the relationship between $[1-2a/3-(1-a)^{2/3}]$ and t, as shown in Fig. 2-78. The correlation coefficients of each fitted straight line are greater than 0.95, and there is a good linear relationship between $[1-2a/3-(1-a)^{2/3}]$ and t, obtaining the reaction rate k_s at each liquid-solid ratio; Using $\ln k_s$ to plot $\ln s$, the results are shown in Fig. 2-79. It can be seen that the liquid-solid ratio reaction order is 0.6185, indicating that increasing the liquid-solid ratio is beneficial for the leaching rate of soluble fluorine in waste cathode carbon blocks from aluminum electrolysis.

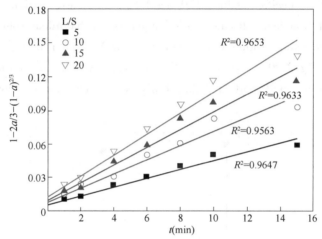

Fig. 2-78 Curves of $[1-2a/3-(1-a)^{2/3}]-t$ under different L/S

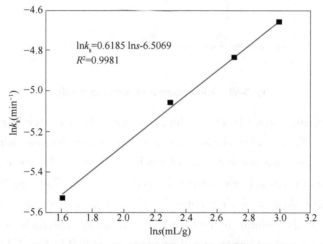

Fig. 2-79 Curve of $\ln k_s$-$\ln s$ under different L/S

From the previous experimental results and kinetic analysis, it can be seen that reducing particle size and increasing liquid-solid ratio can effectively increase leaching rate. Taking into account the actual production situation of the enterprise, reducing production costs, and improving leaching rate, the leaching conditions for soluble fluorine in waste cathode carbon blocks are selected as follows: temperature 85 ℃, particle size 0.058-0.075 mm, liquid-solid ratio 25, selecting 10 g of raw materials for leaching tests (3 times). After 1 h, the liquid is filtered and the filter cake is dried for 2 h. After calculation, the leaching rate of soluble fluorine is as high as 98.9%, and the average concentration of residual soluble fluorine in the leaching residue is 83.53 mg/L, which is lower than the specified value (100 mg/L) in the latest version of the National Hazardous Waste Catalogue. It can be discharged as non hazardous solid waste. Further analysis of the leaching residue was carried out using XRD, and the results are shown in Fig. 2-80. From Fig. 2-80, it can be seen that the majority of the treated leaching residue is carbon, with a small portion being CaF_2, which can be recycled as valuable substances.

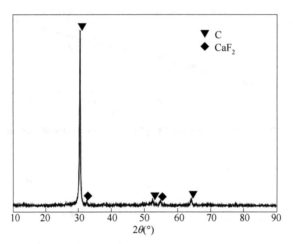

Fig. 2-80 XRD pattern of leaching residue

The waste cathode carbon block in aluminum electrolysis is mainly composed of C, NaF, CaF_2, Na_3AlF_6 and other phases, among which soluble fluorine mainly exists in the form of NaF. The average concentration of soluble fluorine in the waste cathode carbon block determined by ion selective electrode method is about 7500 mg/L. The leaching process of soluble fluorine during water leaching of spent cathode carbon blocks in aluminum electrolysis conforms to the shrinkage core model controlled by diffusion within the solid film layer, with an apparent activation energy of 8.97 kJ/mol. Dynamic analysis shows that temperature has a small impact on the leaching process, and increasing the liquid-solid ratio and reducing the solid particle size can effectively accelerate the internal diffusion rate in the pores of raw material particles. The optimal leaching conditions are: temperature 85 ℃, particle size 0.058 – 0.075 mm, liquid-solid ratio 25 mL/g. After 1 hour of leaching, the soluble fluorine leaching rate in the leaching solution is 98.9%, and the fluorine content in the leaching residue is 83.53 mg/L, which is lower than the safety emission standard of 100 mg/L. The high graphitized solid carbon content in the treated waste cathode carbon block reaches 89.75%, which has high recovery value and can be further recycled as resources.

2.6.2 Leaching of waste side blocks

Part of the waste SiC side blocks were removed and crushed for XRD and chemical composition analysis. The results are shown in Fig. 2-81, and the specific content is shown in Table 2-21. From XRD, it can be seen that the main components of the untreated waste side blocks are SiC, Si_3N_4, NaF, SiO_2, CaF_2, etc. The vast majority of them are SiC and Si_3N_4, with a content of nearly 93%. Fluorides mainly exist in the form of NaF and CaF_2, with a content of about 3.3%. Electrolytes penetrate into the gaps and pores of the side lining material of the electrolytic cell, forming hard and dense waste SiC side blocks that are also difficult to break and grind. The content of waste SiC side blocks varies slightly in different parts.

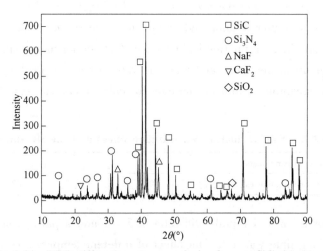

Fig. 2-81 XRD diffraction phase analysis of waste SiC side blocks

Table 2-21 Phase composition of waste sic side blocks

Phase composition	SiC	Si_3N_4	SiO_2	CaF_2	NaF	Others
Content(wt%)	77.72	15.37	1.93	1.74	1.59	1.65

Accurately weigh 10 g of waste SiC side blocks with a particle size of 100-150 mesh, and pour 50 mL of tap water with a liquid-solid ratio of 5 : 1 into a beaker. Weigh 4 times, number 1-4 in sequence, and heat them in a water bath to 30 ℃, 50 ℃, 70 ℃, and 90 ℃, respectively. During the process, cover the pot tightly to prevent evaporation of water and avoid affecting the concentration of soluble fluorine. After reaching the set temperature, add weighed waste SiC side blocks, stir at constant temperature, and measure the soluble fluoride ion content in the leaching solution at regular intervals.

Accurately weigh 10 g of aluminum electrolytic waste SiC side blocks with particle sizes of 50-100 mesh, 100-150 mesh, 150-200 mesh, and below 200 mesh, and measure 50 mL of tap water with a liquid-solid ratio of 5 : 1. Pour them into a beaker, numbered 1-4 in sequence, and heat them in a water bath to 90 ℃. During the process, cover the pot tightly to prevent evaporation of water and avoid affecting the concentration of soluble fluorine.

Accurately weigh 10 g of aluminum electrolytic waste SiC side blocks with a particle size less than 200 mesh, and measure 50 mL, 100 mL, 150 mL, and 200 mL of tap water with a liquid-solid ratio of 3 : 1, 5 : 1, 7 : 1, and 9 : 1. Pour them into a beaker and place them in a water bath to heat to 90 ℃. Cover tightly during the process to prevent the evaporation of water.

Accurately weigh 0.05 g of the raw material and bring it to a 50 mL volumetric flask. Add about 10 mL of buffer solution and bring it to the mark and shake well. Pour into a 150 mL beaker and insert a fluoride ion selective electrode for measurement. The waiting time for each group is the same, and read the value when the reading changes within ±0.5 mV/min. Substitute it into the formula for calculation. The results are shown in Table 2-22.

Table 2-22 Soluble fluorine concentration values in waste SiC side blocks

Number of experiments	1	2	3	Average
Soluble fluorine concentration value (mg/L)	913.5	916.8	883.5	904.6

Under the condition of a liquid-solid ratio of 5 : 1 and a particle size of 100-150 mesh, the leaching time was 1 h. The effect of different temperatures on the leaching rate of soluble fluorine in the raw material was investigated, and the results are shown in Fig. 2-82.

From Fig. 2-82, it can be seen that under the conditions of a liquid-solid ratio of 5 : 1 and a particle size of 100-150 mesh, at a constant temperature, the soluble fluorine leaching amount in the raw material shows a consistent trend over time, gradually

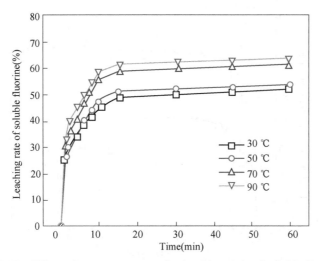

Fig. 2-82 Effect of temperature on the leaching rate of soluble fluorine

increasing with the extension of time, and basically stabilizing at 15 min; For different temperatures, the trend is basically the same, and the leaching amount gradually increases with the increase of temperature. The leaching rate for 15 min at 30 ℃ is 49.3%, and the leaching rate for 15 min at 90 ℃ is 61.7%, with a significant increase. Similar to the effect mechanism of temperature on waste cathode carbon blocks, an increase in temperature will also increase the number of activated molecules in the waste SiC side blocks, accelerating their diffusion rate, thus increasing the leaching rate of soluble fluorine. However, we also see that the leaching rate of soluble fluorine at 70 ℃ is not significantly different from that at 90 ℃. Therefore, when selecting the optimal leaching conditions in the future, comprehensive consideration should be taken.

Under the condition of a temperature of 90 ℃ and a liquid-solid ratio of 5 : 1, the leaching time was 1 h. The effect of different particle sizes on the leaching rate of soluble fluorine in the raw material was investigated, and the results are shown in Fig. 2-83.

From Fig. 2-83, it can be seen that under the conditions of temperature 90 ℃ and liquid-solid ratio 5 : 1, when the particle size decreases from 50-100 mesh to 100-150 mesh, the leaching rate changes significantly. After 15 min of leaching, the leaching rate increases from 40.3% to 58.4%; The particle size continued to decrease to below 200 mesh, and the leaching rate of soluble fluorine reached 68.2% after 15 min of leaching. Overall, as the particle size of the raw material decreases, the leaching rate of soluble fluorine gradually increases, and with the decrease of particle size, the increase in leaching rate of soluble fluorine is more significant. For solid particles. The area of the particle reaction interface increases with the decrease of particle size, and the larger the

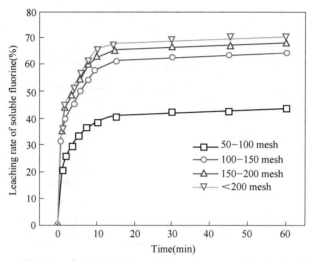

Fig. 2-83 Effect of particle size on the leaching rate of soluble fluoride

reaction area, the more reactions occur with the leaching solution per unit time, and the faster the leaching rate.

Under the conditions of temperature 90 ℃ and particle size less than 200 mesh, leaching for 1 h was conducted to investigate the effect of different liquid-solid ratios on the leaching rate of soluble fluorine in the raw material. The results are shown in Fig. 2-84.

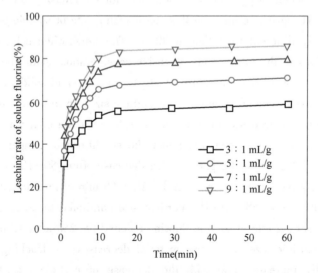

Fig. 2-84 Effect of liquid-solid ratio on soluble fluorine leaching rate

From Fig. 2-84, it can be seen that as the liquid-solid ratio gradually increases, the leaching rate of soluble fluorine in the waste SiC side block rapidly increases. When the liquid-solid ratio increases from 3 : 1 to 9 : 1, after 15 min of leaching, the leaching rate of soluble fluorine increases from 54.6% to 84.7%. The leaching curve in the figure can be clearly divided into two stages: the first stage is that the leaching time is within 15 min. The second stage is the later stage of leaching, with a leaching time greater than 15 min. Due to the fact that most of the exposed soluble fluorine has been leached, it is difficult for the fluorine in the particles to be leached. Continuing to extend the leaching time, it was found that the increase in leaching rate is relatively small.

When the limiting link for the leaching and diffusion of waste SiC side blocks is controlled by chemical reactions, the leaching rate of soluble fluorine may increase several times when the temperature increases by 10 ℃. However, based on the experimental results in Fig. 2-82, it is speculated that the diffusion limiting link for soluble fluorine in this experiment should not be a chemical reaction. To further verify, the leaching rates of soluble fluorine obtained at different temperatures were determined a the linear fitting of time t indicates that mixed diffusion control may be suitable for the leaching of soluble fluorine in waste SiC side blocks from aluminum electrolysis, and the mixed diffusion core model is chosen for calculation.

Under the conditions of a liquid-solid ratio of 5 : 1 and a particle size of 100–150 mesh, the experimental results in Fig. 2-82 were calculated according to $[\ln(1-a)]/3 - 1 + (1-a)^{-1/3}$ of t is plotted, and the results are shown in Fig. 2-85. From the figure, it can be seen that $[\ln(1-a)]/3 - 1 + (1-a)^{-1/3}$ has a good linear relationship with t, and

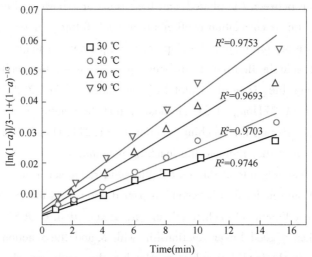

Fig. 2-85 Relationship between $[\ln(1-a)]/3 - 1 + (1-a)^{-1/3}$ and t at different temperatures

the correlation coefficient (R^2) at all temperatures is above 0.955, indicating a good linear relationship. The slope of the fitting equation is the leaching rate of soluble fluoride at each temperature, which is then substituted into the Arrhenius formula to obtain the apparent activation energy. The results are shown in Fig. 2-86. Furthermore, the apparent activation energy $E_a = 13.67$ kJ/mol for the soluble fluorine water leaching reaction of aluminum electrolysis waste SiC side blocks can be calculated.

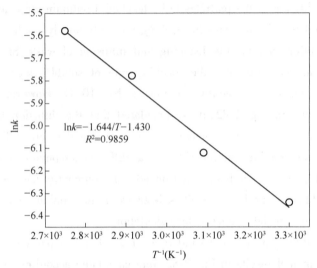

Fig. 2-86 Relationship between ln$k$ and $1/T$

Calculate the leaching rate at different particle sizes according to $[\ln(1-a)]/3 - 1 + (1-a)^{-1/3}$ of the relationship between t is plotted, and the results are shown in Fig. 2-87. As shown in the figure, the linear correlation coefficients of each fitting equation are above 0.95, with a high degree of correlation. The slope of the equation is calculated as the reaction rate constant of soluble fluorine at different particle sizes. In combination with the Arrhenius equation, lnk_d is used to plot lnd_0, as shown in Fig. 2-88. The fitted linear equation is ln$k_d = -1.271\ln d_0 + 0.316$, indicating that the reaction order of the waste SiC side block particle size on the leaching reaction is -1.271. This indicates that reducing the particle size can better promote the leaching reaction.

Calculate the leaching rate at different liquid-solid ratios according to $[\ln(1-a)]/3 - 1 + (1-a)^{-1/3}$ of the relationship between t is plotted, and the results are shown in Fig. 2-89. The correlation coefficients of each fitted line are greater than 0.9653, $[\ln(1-a)]/3 - 1 + (1-a)^{-1/3}$ has a good linear relationship with t, and the reaction rate k_e at each liquid-solid ratio is obtained; Using lnk_e to plot ln$e$, the results are shown in Fig. 2-90. The fitted linear equation is ln$k_e = 1.41\ln e - 8.13$, and the reaction order of liquid-solid

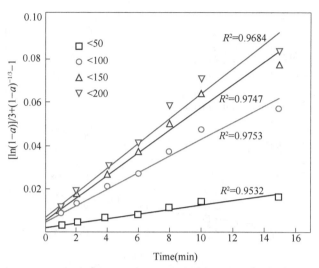

Fig. 2-87 Relationship between $[\ln(1-a)]/3-1+(1-a)^{-1/3}$ and t at different granularity

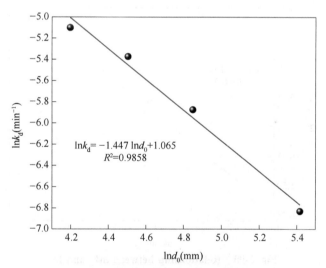

Fig. 2-88 Relationship between $\ln k_d$ and $\ln d_0$ at different granularities

ratio is 1.41. This indicates that increasing the liquid-solid ratio is beneficial for improving the leaching rate of soluble fluorine in aluminum electrolysis waste SiC side blocks.

Based on the previous experimental results and kinetic analysis, it can be concluded that reducing particle size and increasing liquid-solid ratio can effectively increase the

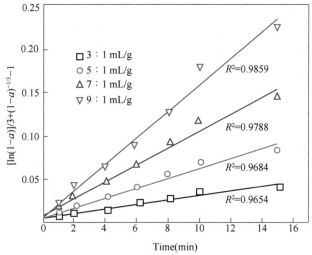

Fig. 2-89 Relationship between $[\ln(1-a)]/3-1+(1-a)^{-1/3}$ and t under different liquid-solid ratios

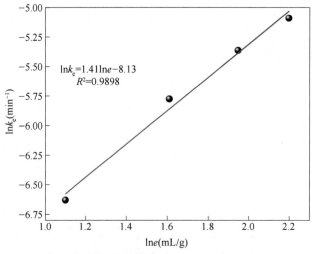

Fig. 2-90 Relationship between $\ln k_e$ and $\ln e$ under different liquid-solid ratios

leaching rate of soluble fluorine, taking into account factors such as reducing production costs and improving leaching rate. The optimal leaching conditions for soluble fluorine in waste SiC side blocks are selected as follows: temperature 80 ℃, particle size less than 200 mesh, liquid-solid ratio 11 ∶ 1, 10 g of raw material for leaching test, 1 h later, the solution is filtered, and the filter cake is dried for 2 h. Measure the leaching solution multiple times and finally obtain the average value. After calculation, the average leaching rate of soluble fluorine reached 99.2%. The detection and analysis of soluble fluorine content in the leaching residue showed an average concentration of 24.12 mg/L, which is far lower than the specified value (100 mg/L) in the latest version of the National Hazardous Waste List. This means that the soluble fluorine in the SiC side block of aluminum electrolysis waste has reached harmless treatment. Further analysis of the leaching residue was carried out using an X-ray diffractometer, and the results are shown in Fig. 2-91. From the figure, it can be seen that NaF has been basically leached from the waste SiC side block of aluminum electrolysis after the experiment.

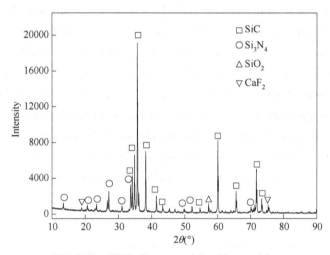

Fig. 2-91 XRD diagram of leaching residue

The waste SiC side blocks from aluminum electrolysis are mainly composed of phases such as SiC, Si_3N_4, NaF, SiO_2, CaF_2, etc. Among them, soluble fluorine mainly exists in the form of NaF. The average concentration of soluble fluorine in the waste SiC side blocks determined by ion selective electrode method is about 1939.4 mg/L. The limiting step in the leaching process of soluble fluorine from waste SiC side blocks in aluminum electrolysis is mixed diffusion control, with an apparent activation energy of 11.89 kJ/mol. Dynamic analysis shows that increasing temperature has little effect on the leaching process. Increasing the liquid-solid ratio and reducing the solid particle size can effectively accelerate the internal diffusion rate in the pores of raw material particles. The optimal leaching conditions are: temperature of 80 ℃, particle size less than 200 mesh, liquid-solid ratio of 11 ∶ 1, leaching time of 1 h. After measurement, the leaching rate of soluble fluorine is 99.2%, and the remaining soluble fluorine concentration in the leaching residue is 24.12 mg/L, which is far below the safety discharge standard of 100 mg/L. The XRD detection analysis of the leaching residue shows that the residue mainly contains substances such as SiC, Si_3N_4, CaF_2, etc., and the remaining soluble fluorine content in the leaching residue is also controlled within the limit range of hazardous solid waste. It can be discharged or recycled as valuable material.

2.6.3 Leaching of waste refractory materials

Part of the waste refractory material was removed and crushed for XRD and chemical composition analysis. The results are shown in Fig. 2-92, and the element content is shown in Table 2-23. The refractory material in the lower part of the aluminum electrolytic cell reacts with the permeated electrolyte to produce different products depending on the ratio of silicon oxide to alumina. In the production process of aluminum electrolysis enterprises in China, dry refractory materials are generally used, with SiO_2 content generally around 55% and Al_2O_3 content around 35%. After the reaction, $NaAlSiO_4$ (nepheline) is mainly generated. Based on the XRD diagram, it can be seen that the main components in the waste refractory material are $NaAlSiO_4$, SiO_2, aluminum sodium oxide, and fluoride salts. Fluorine mainly exists in the forms of NaF and CaF_2. The content of Na and F in the anti-seepage material increases sharply after the electrolytic cell has been working for a period of time, indicating that the electrolyte has penetrated into the anti-seepage material at this time, and the aluminum electrolytic cell needs to be overhauled.

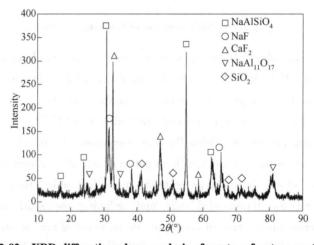

Fig. 2-92 XRD diffraction phase analysis of waste refractory materials

Table 2-23 Chemical composition of waste refractory materials

Chemical composition	Al	F	Na	O	Si	Ca	K	Fe
Content(wt%)	29.02	28.15	17.11	11.94	6.71	2.51	2.08	1.20

Accurately weigh 10 g of waste refractory material with a particle size of 100-150 mesh, and pour 50 mL of tap water with a liquid-solid ratio of 5 : 1 into a beaker. Weigh 4 times, number 1-4 in sequence, and heat them in a water bath to 30 ℃, 50 ℃, 70 ℃, and 90 ℃ respectively. During the process, cover the pot tightly to prevent evaporation of water and avoid affecting the concentration of soluble fluorine. After reaching the set temperature, add weighed waste refractory material, stir at constant temperature, and measure the soluble fluorine content in the leaching solution at regular intervals.

Accurately weigh 10 g of aluminum electrolytic waste refractory materials with particle sizes of 50-100 mesh, 100-150 mesh, 150-200 mesh, and below 200 mesh, and measure 50 mL of tap water with a liquid-solid ratio of 5 : 1. Pour them into a beaker, numbered 1-4 in sequence, and heat them in a water bath to 90 ℃. During the process, cover the pot tightly to prevent the evaporation of water and avoid affecting the concentration of soluble fluorine.

Accurately weigh 10 g of aluminum electrolytic waste refractory materials with particle size less than 200 mesh, and measure 50 mL, 100 mL, 150 mL, and 200 mL of tap water with a liquid-solid ratio of 4 : 1, 5 : 1, 6 : 1, and 7 : 1. Pour them into a beaker and place them in a water bath to heat to 90 ℃. Cover tightly during the process to prevent the evaporation of water.

Accurately weigh 0.05 g of aluminum electrolytic waste refractory material, add it to a 50 mL volumetric flask, and add about 10 mL of buffer solution. Shake well after reaching a constant volume. Pour into a 150 mL beaker and insert a fluoride ion composite electrode for measurement. The waiting time for each group of measurements is the same. When the reading changes within ±0.5 mV/min, read the value and bring it into the formula for calculation. The results are shown in Table 2-24.

Table 2-24　Soluble fluorine concentration values in waste refractory materials

Number of experiments	1	2	3	Average
Soluble fluorine concentration value(mg/L)	1901.7	1938.8	1977.7	1939.4

Under the condition of a liquid-solid ratio of 5 : 1 and a particle size of 100-150 mesh, the leaching time was 1 h. The effect of different temperatures on the leaching rate of soluble fluorine in the raw material was investigated, and the results are shown in Fig. 2-93.

From Fig. 2-93, it can be seen that under the condition of a liquid-solid ratio of 5 : 1 and a particle size of 100-150 mesh, the soluble fluorine in the raw material increases with the increase of temperature for 1 h. The change is more intense in the first 15 min,

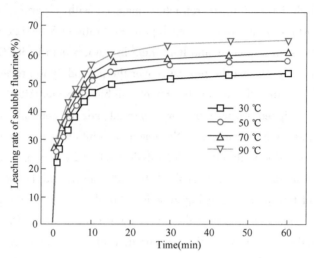

Fig. 2-93　Effect of temperature on the leaching rate of soluble fluorine

and tends to stabilize after 15 min. From the graph, it can be seen that the effect of temperature on its leaching rate is quite obvious. Temperature is the main control parameter for molecular thermal motion. The increase in temperature will intensify the thermal movement of molecules, thereby increasing the leaching rate. As the temperature increases, the leaching rate of soluble fluorine in the waste cathode gradually increases. When the temperature rises from 30 ℃ to 90 ℃, after 15 min of leaching, the leaching rate of soluble fluorine increases from 49.2% to nearly 59%. However, the impact of temperature increase on the leaching rate is not particularly significant.

Under the condition of a temperature of 90 ℃ and a liquid-solid ratio of 5 : 1, the leaching time was 1 h. The effect of different particle sizes on the leaching rate of soluble fluorine in the raw material was investigated, and the results are shown in Fig. 2-94.

From Fig. 2-94, it can be seen that under the conditions of temperature 90 ℃ and liquid-solid ratio 5 : 1, when the particle size decreases from 50-100 mesh to 100-150 mesh, the change is particularly significant, and the soluble fluorine leaching rate increases from 40% to about 60%; Continue to reduce to 150-200 mesh, with little change in leaching rate; The soluble fluorine content reached a maximum of 68.4% after 15 min of leaching with a particle size of 200 mesh or less, and the subsequent leaching amount of soluble fluorine did not change significantly with time. Overall, as the particle size of the raw material decreases, the leaching rate of soluble fluorine gradually increases, and the trend of increasing the leaching rate of soluble fluorine with the decrease of particle size is more obvious. For solid particles, the area of the particle

2.6 Hydrometallurgy disposal technology · 165 ·

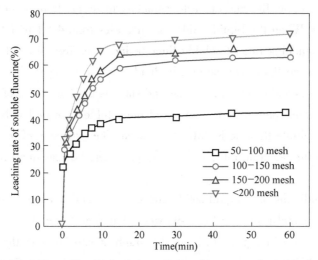

Fig. 2-94 Effect of particle size on the leaching rate of soluble fluoride

reaction interface increases with the decrease of particle size. The larger the reaction area, the more reactions occur with the leaching solution per unit time, and the faster the leaching rate.

Under the conditions of temperature 90 ℃ and particle size less than 200 mesh, leaching for 1 h was conducted to investigate the effect of different liquid-solid ratios on the leaching rate of soluble fluorine in the raw material. The results are shown in Fig. 2-95.

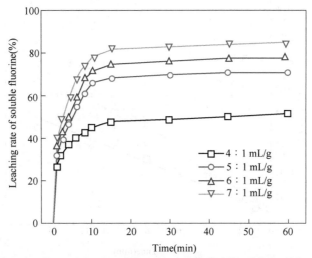

Fig. 2-95 Effect of liquid-solid ratio on soluble fluorine leaching rate

From Fig. 2-95, it can be seen that as the liquid-solid ratio gradually increases, the leaching rate of soluble fluorine in waste refractory materials from aluminum electrolysis rapidly increases. When the liquid-solid ratio increases from 4∶1 to 7∶1, after 15 min of leaching, the leaching rate of soluble fluorine increases from 48.6% to 82.7%. The leaching curve in the figure can be clearly divided into two stages: the first stage is that the leaching time is within 15 min. The second stage is the later stage of leaching, with a leaching time greater than 15 min. As most of the soluble fluorine has been leached, the remaining little soluble fluorine is difficult to leach. The extension of reaction time has a small impact on the leaching rate, but as time prolongs, the leaching rate still increases slightly.

Under the conditions of a liquid-solid ratio of 5∶1 and a particle size of 100–150 mesh, the experimental results in Fig. 2-93 were plotted as $1-2a/3-(1-a)^{2/3}$ against t. The results are shown in Fig. 2-96. From the graph, it can be seen that $1-2a/3-(1-a)^{2/3}$ has a good linear relationship with t, and the correlation coefficient (R^2) at each temperature is above 0.960, indicating a good linear relationship. The slope of the fitted straight line is the leaching rate at different temperatures, and the apparent activation energy is calculated by combining it with the Arrhenius formula. The results are shown in Fig. 2-97. Furthermore, the apparent activation energy $E_a = 8.65$ kJ/mol for the soluble fluorine water leaching reaction of aluminum electrolysis waste cathode carbon blocks can be calculated. Based on the apparent activation energy, it can be inferred that the leaching process follows the internal diffusion control model.

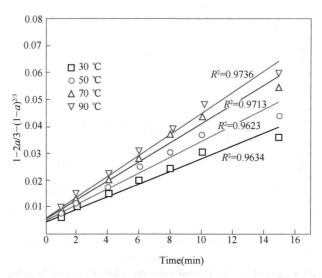

Fig. 2-96　Relationship between $1-2a/3-(1-a)^{2/3}$ and t at different temperatures

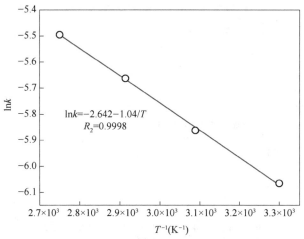

Fig. 2-97 Relationship between ln$k$ and 1/T

Based on the leaching rate results at different particle sizes mentioned above, the relationship between $1-2a/3-(1-a)^{2/3}$ and t is plotted by substituting the equation, as shown in Fig. 2-98. From the figure, it can be seen that the correlation coefficient values of each fitted straight line are relatively large ($\geqslant 0.9603$), and the fitting results are good. The reaction rate constant k_d at each particle size is obtained; Combining the Arrhenius equation, lnk_d is used to plot lnd_0. The results are shown in Fig. 2-99. The fitted linear equation is ln$k_d = -0.416 \ln d_0 - 3.575$, indicating that the reaction order of the waste refractory material particle size on the leaching reaction is -0.416. This indicates that reducing the particle size can better promote the leaching reaction.

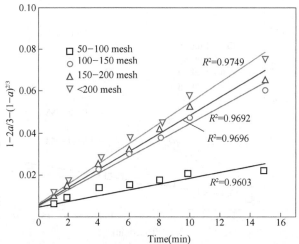

Fig. 2-98 Relationship between $1-2a/3-(1-a)^{2/3}$
and time at different granularities

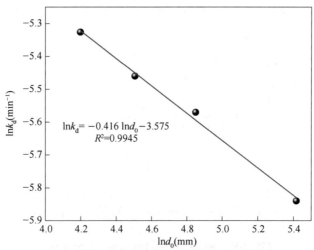

Fig. 2-99　Relationship between $\ln k_d$ and $\ln d_0$ at different granularities

Plot the leaching rates under different liquid-solid ratios according to the relationship between $1-2a/3-(1-a)^{2/3}$ and t, as shown in Fig. 2-100. The correlation coefficient of each fitted straight line is greater than 0.9597, and $1-2a/3-(1-a)^{2/3}$ has a good linear relationship with t, obtaining the reaction rate k_s at each liquid-solid ratio; Using $\ln k_s$ to plot $\ln s$, the results are shown in Fig. 2-101. The fitted linear equation is $\ln k_s = 2.324 \ln s - 9.153$, and the liquid-solid ratio reaction order is 2.324. This indicates that increasing the liquid-solid ratio is beneficial for improving the leaching rate of soluble fluorine in aluminum electrolysis waste refractory materials.

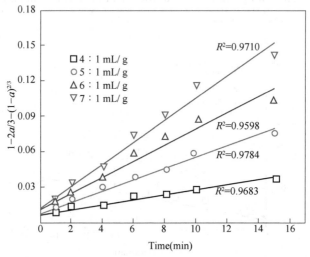

Fig. 2-100　Relationship between $1-2a/3-(1-a)^{2/3}$ and time t under different liquid-solid ratios

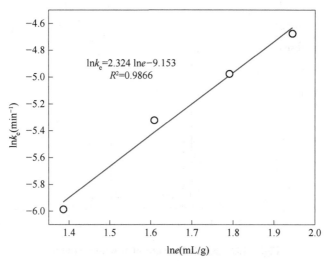

Fig. 2-101 Relationship between ln$s$ and lnk_s under different liquid solid ratios

Based on the results of the previous leaching experiment and kinetic analysis, it can be concluded that reducing particle size and increasing the liquid-solid ratio can effectively increase the leaching rate of soluble fluorine in aluminum electrolysis waste refractory materials. Taking into account production costs and the actual situation of the enterprise, the optimal leaching conditions for soluble fluorine in waste refractory materials are selected as follows: Temperature 80 ℃, particle size less than 200 mesh, liquid-solid ratio 8 : 1, take 10 g of raw materials for leaching test, and filter the solution after 1 h, The filter cake is dried for 2 h, and the soluble fluorine leaching rate and the phase composition of the leaching residue in the leaching solution are tested separately. The leaching solution can be obtained multiple times to obtain the average value. After measuring with the fluoride ion selective electrode method, the leaching rate of soluble fluorine was calculated to reach 99.1%. The detection and analysis of soluble fluorine content in the leaching residue showed an average concentration of 89.05 mg/L, which is much lower than the specified value (100 mg/L) in the latest version of the National Hazardous Waste List. This means that the soluble fluorine in aluminum electrolysis waste refractory materials has been treated harmless. Further analysis of the leaching residue was carried out using an X-ray diffractometer, and the results are shown in Fig. 2-102. From the figure, it can be seen that NaF was not detected in the leached aluminum electrolysis waste refractory material, indicating that most of the NaF was dissolved in the leaching solution.

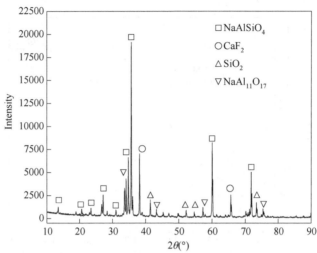

Fig. 2-102　XRD diagram of leaching residue

The waste refractory materials from aluminum electrolysis mainly include $NaAlSiO_4$, SiO_2, aluminum sodium oxide, and fluoride salts, among which soluble fluorine mainly exists in the form of NaF. The average concentration of soluble fluorine in the waste refractory materials determined by ion selective electrode method is about 904.6 mg/L. The leaching experiment and kinetic analysis of waste refractory materials show that the leaching process of soluble fluorine conforms to the diffusion diffusion control model within the solid film layer, and its apparent activation energy is 8.65 kJ/mol; Increasing temperature has little effect on the leaching process, while reducing particle size and increasing liquid-solid ratio can effectively accelerate the leaching rate of soluble fluoride in raw materials. The optimal leaching conditions are: temperature of 80 ℃, particle size less than 200 mesh, liquid-solid ratio of 8∶1, leaching time of 1 h. After measurement, the leaching rate of soluble fluorine is 99.1%, and the remaining soluble fluorine concentration in the leaching residue is 89.05 mg/L, which is lower than the safety discharge standard of 100 mg/L. The XRD analysis of the leaching residue shows that the leaching residue of the waste refractory material after leaching mainly contains substances such as $NaAlSiO_4$, SiO_2, and aluminum sodium oxide. The soluble fluorine content in the leaching residue is also controlled within the limit of hazardous solid waste, which can be discharged or recycled as valuable substances.

2.7　Analysis on thermal behavior of fluorides and cyanides by TG/DSC-MS & ECSA

This part adopts the TG/DSC-MS analysis combining a novel data-processing method of ECSA, to basically study the characteristics of release and transformation of fluorides and

2.7 Analysis on thermal behavior of fluorides and cyanides by TG/DSC-MS & ECSA

cyanides during heat treatment of the spent cathode carbon block (SCCB). All the experiments were conducted at 10 K/min heating rate and under Ar or Ar-O_2 atmospheres. The release of fluorides and transformation of cyanides were qualitatively and quantitatively obtained. The results indicate that the release of fluorides was just a steady but slow phase transition process under both Ar and Ar-O_2 atmospheres, and this process can be radically accelerated when the carbon material was burnt. The cyanides were effectively decomposed at the high temperature and at Ar-O_2 atmosphere. Quantitative calculation indicates that around three quarters of the cyanides were converted to the N_2, and the rest was mainly converted to the NO. There was a relatively small quantity of NO_2 and HCN produced. The flue gas composition under Ar and Ar-O_2 atmospheres were also analyzed and compared. The flue gas under Ar-O_2 atmosphere mainly consists of CO_2, N_2, NO, NO_2, HCN; while that under Ar atmosphere mainly consists of CO_2 and NO.

The primary aluminum is mainly produced in the cryolite-alumina molten salt electrolysis process, which is conducted in cells about 1-1.5 m in height, 5-15 m in length, and 3-5 m in width at present. The cell is filled with the molten mixture of alumina, cryolite and the other fluoride salts between two electrodes, held in temperatures around 950 ℃. The carbon anode hangs above the molten bath while the cathode carbon block is arranged regularly on the bottom of the cell connected to steel bars, playing functions of conducting current and bearing the high temperature molten mixture.

The cathode carbon block used in aluminum reduction cell is often made from electric calcined anthracite, graphite scrap and coal tar pitch through mixing, shaping, roasting and graphitization[119]. At present, most of the cathode carbon blocks used in large aluminum reduction cell is graphite cathode carbon blocks with graphite content of 30% or 50%, of which graphitized carbon accounts for about 60% - 70%. However, the cathode carbon blocks will be damaged after the aluminum reduction cell is used for a certain period of time, generally 4-6 a. The continuous chemical corrosion and physical erosion from the high temperature molten metal and salts wear the blocks and make the cracks formed during its lifetime. Consequently, molten aluminum, fluorides would penetrate into the cathode block through the cracks slowly, and fluorides mixed with Al-Fe alloy layer, Al, Na and other metals have been finally formed in the voids of blocks. As well, secondary products can be generated by reactions involving chemical species adsorbed onto the cathode carbon blocks. For instance, the nitrogen from the air can react with the carbon of the cathode block and the sodium(Na) of the molten salts to form sodium cyanide NaCN as follows,

$$2Na + 2C + N_2 \longrightarrow 2NaCN \tag{2-66}$$

The oxygen from the air can react with the penetrated Na and Al_2O_3 to form β-Alumina ($Na_2O \cdot 11Al_2O_3$) as follows[120],

$$2Na + 0.5O_2 + 11Al_2O_3 \longrightarrow 2Na_2O \cdot 11Al_2O_3 \qquad (2-67)$$

The penetrated Al liquid, fluorides can react with the steel bars to form Al-Fe alloy as follows,

$$2Na_3AlF_6 + 3Na + xFe \longrightarrow AlFe_x + 6NaF \qquad (2-68)$$

$$Al + xFe \longrightarrow AlFe_x \qquad (2-69)$$

When overhauling the reduction cell, the damaged cathode carbon blocks need to be replaced and discarded, which is often given a name of "spent cathode carbon block (SCCB)". It is estimated that for every tonne of primary aluminum produced 8-10 kg SCCB is generated, and the total amount of SCCB generated is around 0.85 Mt/a in the world. The SCCB contains highly valuable graphitized carbon material. However, due to containing a large amount of soluble fluorides (in the level of 2000-4000 mg/L) and a certain amount of toxic cyanides (in the level of 10-20 mg/L), the SCCB is listed by various environmental bodies as a hazardous material. Long-term storage and unreasonable disposal of the SCCB could cause diffusion of the soluble fluorides and cyanides into the soil and groundwater, and further the contamination of human and animal food supplies. Disposal of such huge amounts of hazardous SCCB has been becoming the utmost important issue for aluminum producers.

The prime target of treating the hazardous SCCB is to remove the fluorides and cyanides clearly. Numerous potential treatment technologies had been developed over the last decades. The hydrometallurgy and pyrometallurgy processes are two types of treatment strategies. Firstly, a representative of hydrometallurgy is LCL & L by Rio Tinto Aluminum in which the cyanides are destroyed by oxidation and fluorides are fixed by the addition of limestone and formation of CaF_2. A domestic aluminum plant had built a line for treating waste cathode carbon blocks, adopting the grinding-flotation-acid leaching-evaporation process. Although the hydrometallurgy process realizes to some extent the removal of fluorides and cyanides, it is still limited by the issues such as complicated flow, high investment, low-quality products and serious pollution. It should be pointed out that the hydrometallurgy is difficult to completely separate carbon powder from fluoride salts. In addition, a large number of HF and HCN gases would be precipitated during the treatment process, causing corrosion damage to equipment and secondary pollution of water. Alternatively, the pyrometallurgy process is thought to be a more promising treatment of the SCCB. The pyrometallurgy process often operates in reactors, such as rotating drum, fluidized bed, even a molten slag bath, under oxidizing atmosphere at medium or high temperatures. The SCCB is first crushed, then burnt so as to make the cyanides decomposed or destroyed, finally make fluorides volatilized or fixed at high temperature before landfill. The SCCB is also burnt by adding different additives such as pet coke, pitch etc. to produce the self-baking anode, or by lignite, limestone to co-produce cement. Such traditional pyrometallurgy process could not recover effectively the valuable carbon materials and produce new solid waste residue. Therefore, there is strong incentive to recover simultaneously the graphite carbon material and fluorides. Recently, heat treatments at high temperature or even ultra-high temperature are gradually rising and gain much more attentions. For instance, a joint temperature-vacuum controlling process for treating SCCB is proposed by our research group and is advancing towards the industrialization.

Actually, no matter the traditional pyrometallurgy process or the high temperature heat treating process, which are uniformly referred to as "heat-treating process" in this paper, most of cyanides and partial fluorides must be transferred into the flue gas. Qualitative and quantitative analysis on the thermal behavior such as release and transformation characteristics of fluorides and cyanides cannot be well done now and the relevant basic data can rarely be found in the literature. It will get the designer into troubles of determining the operating conditions such as temperature, residence time if the quantitative transformation data cannot be obtained; and also of choosing the purification devices like denitrification device if the composition and quantities of the flue gas, especially the pollution gases such as HF, NO_x, is still unclear. Therefore there will be of great significance to clearly understand the release and transformation characteristics of fluorides and cyanides for heat-treating SCCB, and to provide a qualitative and quantitative data of gases composition released in the flue gas.

The objective of this article is to basically measure and analyze the characteristics of release and transformation of fluorides and cyanides during heat-treating SCCB, to obtain qualitatively the composition of flue gas, and to determine quantitatively the content of pollution gases in the flue gas by using the thermogravimetry-Differential Scanning Calorimeter-mass spectrometry (TG/DSC-MS) analysis combining a novel data-processing method of equivalent characteristic spectrum analysis (ECSA).

2.7.1 Materials and characterization

The waste cathode carbon blocks used in this study were taken from an aluminum electrolysis cell operated by a smelter in Ningxia Hui Autonomous Region of China for over 2000 days. The raw spent cathode carbon block was broken down by using a jaw crusher, then sieved to obtain two types of particle size 1 mm (label as S-0) and 100 μm (label as S-1). In addition, the raw material obtained was dried in an oven at temperature 80 ℃ for 24 h to eliminate the moisture for subsequent experiments and characterization tests.

Observation and analysis by optical microscope (OM) and scanning electron microscope (SEM) show in Fig. 2-103 that the surface of SCCB is eroded to varying degrees, forming non-uniform but dense erosion pits/holes into which fluorides and Al-Fe alloy layer, Al, Na and other metals are embedded. SEM clearly shows that the fluorides (bright substances as shown in Fig. 2-103 (b)) stick the carbon particles firmly. Combining the EDS analysis in Table 2-25 gives notably an uneven distribution of elements at different positions measured, but with the main elements of C, O, F and Na. It was inferred that the carbon particles and fluorides are more likely to be physical adheren and combination. Nevertheless, a simple physical process cannot effectively separate the fluorides from the carbon materials due to combination in the microscale.

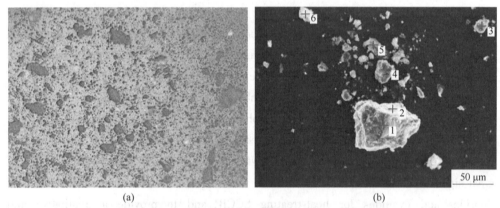

(a) (b)

Fig. 2-103 Morphology of SCCB used in this study
by OM analysis(a) and SEM analysis(b)

In addition, Table 2-25 summarizes the chemical composition of the SCCB used in this study. It indicates the SCCB mainly consists of carbon and fluorides, accounting totally for around 91%. The fluorides mainly exist in the crystalline phases of the NaF (with a melting point of 993 ℃), Na_3AlF_6 (with a melting point of 1009 ℃), and CaF_2 (with a melting point of 1423 ℃). Except for a high level of soluble fluoride content of 2342 mg/L, there is an amount of cyanide of 15.6 mg/L, i.e. 0.03%.

Table 2-25 EDS analysis and chemical composition of SCCB used in this study

(%)

(1) EDS analysis on different positions of SCCB measured										
Elements	C	O	F	Na	Al	Ca	Fe	K	Mg	Si
P1	83.70	8.57	0.75	6.97						
P2	67.97	24.05	4.36	3.62						

Continued Table 2-25

(1) EDS analysis on different positions of SCCB measured										
Elements	C	O	F	Na	Al	Ca	Fe	K	Mg	Si
P3	96.15			2.19			1.66			
P4	71.81		18.62	4.76	1.56	3.17				
P5	84.12	13.69		2.19						
P6	7.55	11.87	36.14	34.37	7.78			0.92	0.48	0.88
(2) Chemical composition of SCCB used as raw material										

No.	Chemical composition	SCCB	Melting point(℃)
1	Fixed carbon[①](%)	74.08	
2	NaF(%)	9.72	993
3	Na_3AlF_6(%)	4.86	1009
4	CaF_2(%)	2.43	1423
5	Other content(%)	8.36	
6	Soluble fluorine content[②](mg/L)	2342	
7	Cyanide content[②](mg/L)	15.6	

①The Fixed carbon content measured by sample dissolution method.
②Determined acoording to GB 5085.3—2007.

2.7.2 TG/DSC-MS measurements and analysis by ECSA

The measurements were conducted on a Netzsch STA 449F3 thermoanalyser coupled with a Quadruple Mass Spectrometer (QMS) 403C Aëolos. The device operated without sample in inert atmosphere (high purity analytical grade dry Ar, 99.99%) every time before measurements to eliminate the memory effects.

The measurements were conducted for both samples under inert and combustion atmospheres. The measurements under inert atmosphere were carried out at 10 K/min heating rate under high purity analytical grade argon dry gas (Ar, 99.99%) with a flow rate of 100 mL/min. While the combustion measurements were carried out under 65 mL/min Ar and 5 mL/min oxygen (O_2). All samples were weighed around 20 mg in Al_2O_3 crucibles and heated up to maximum 1100 ℃ from room temperature, with a controlled temperature program. Each set of measurements were repeated twice to verify the results. The detail conditions of TG/DSC-MS measurements are shown in Table 2-26.

Table 2-26　Conditions of the TG/DSC-MS measurements

Samples	Test and analysis	Weight (mg)	Heating rate (K/min)	Ar flow rate (mL/min)	O_2 flow rate (mL/min)
S-0	TG/DSC-MS & ECSA	20.0	10	100	0
S-0	TG/DSC-MS & ECSA	20.0	10	65	5
S-1	TG/DSC-MS & ECSA	20.2	10	100	0
S-1	TG/DSC-MS & ECSA	20.0	10	65	5

Note that relative to a rough and ready measurement, a novel method of equivalent characteristic spectrum analysis (ECSA) was used to deal with data, i. e., the ion current of mass spectrum that measured directly by the instrument. The principle and operation of ECSA based on the TG/DSC-MS measurement can be found in the literatures. It has been proven more than once, that the ECSA can effectively eliminate the ion current overlap, mass discrimination of MS and the temperature-dependent effect of TG, and can also accurately determine the mass flow rate of various components in the evolved gas.

2.7.3 Thermal behavior of fluorides under Ar and Ar-O_2 atmospheres

At first, the analysis focuses on the thermal behavior of treating SCCB with particle size of 1 mm, i. e. sample S-0, under Ar atmosphere and Ar-O_2 atmospheres. Under Ar-O_2 atmosphere, Fig. 104(a)-(c) show curves of mass loss, DTG, and DSC obtained from measurements, mass flow rate of gases including CO_2, O_2, H_2O, and the amount of burnt carbon calculated by ECSA. In Fig. 2-104(a), the mass loss occurs at the beginning of 100 min (corresponding around 750 ℃), lasts until the end of measurement (165 min, 1100 ℃). While the real-time change curves of mass flow rates of CO_2, O_2 in Fig. 2-104 (b) show that the production of CO_2 and consumption of O_2 due to the brunt of carbon, occurs only at the range of 100-140 min. This implies that the mass loss after 140 min mainly results from the volatilization of fluorides. For verifying this result and determining quantitatively the fluorides volatilization rate, the mass loss resulted from C elements was deduced from the CO_2, COS and HCN by ECSA, as "Cal Mass_C" shown in Fig. 2-104(c), which was used to compare with the measured mass loss of sample. Actually, the volatilization of fluorides starts at around 135 min (close to 1000 ℃), slightly ahead of the burnout of carbon. Combining the physical properties of fluorides including the NaF (melting point 993 ℃ at 1 atm), and the Na_3AlF_6 (melting point 1009 ℃ at 1 atm), it can be concluded that the fluorides can only be volatilized when the temperature is beyond their respective melting point and is almost not affected by the carbon combustion. The volatilization of fluorides is just a phase transition process, with an averaged DTG of 0.20%/min, i. e., 0.0408 mg/min.

The thermal behavior of treating sample S-0 under Ar atmosphere, as shown in Fig. 2-105(a)-(c), was analyzed to further clarify the volatilization of fluorides. As shown from the mass loss curve in Fig. 2-105(a), there is only 7% mass loss under Ar atmosphere compared to the around 90% mass loss under combustion process given the same reaction time. Such 7% mass loss should be resulting from both the volatilization of fluorides and evolution of CO_2 produced between C and O elements in SCCB particle. Nevertheless,

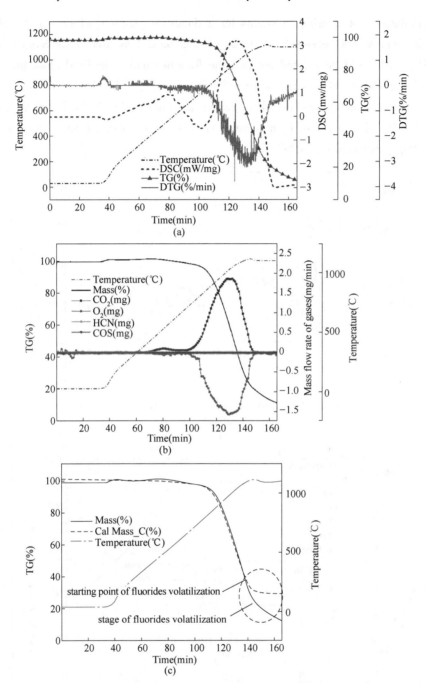

Fig. 2-104 Thermal behavior for 1 mm SCCB (sample S-0) under Ar-O$_2$ atmosphere

(a) Mass loss, DTG, and DSC curves; (b) Mass flow rate of gases including CO_2, O_2, H_2O calculated by ECSA;
(c) Comparison between the mass loss on C element calculated by ECSA and mass loss from measurement

the volatilization of fluorides accounts for a dominant position, which can be affirmed from the comparison between "Cal Mass_C" and mass loss from measurement in Fig. 2-105(c). It should be pointed out that the fluorides only volatilized close to 1000 ℃ while the production of CO_2 can occur at lower temperature. The DSC and DTG curves in Fig. 2-105(a) show that the fluorides volatilization rate has levelled off, with an averaged DTG of 0.022%/min (0.0044 mg/min), after a quick volatilization (corresponding to sharp drop peaks) close to 1000 ℃.

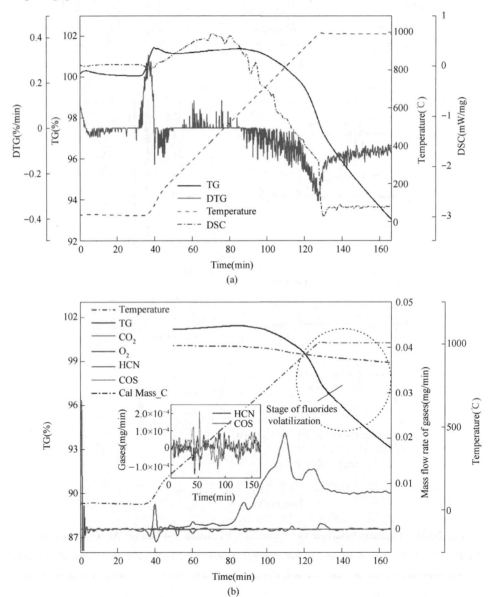

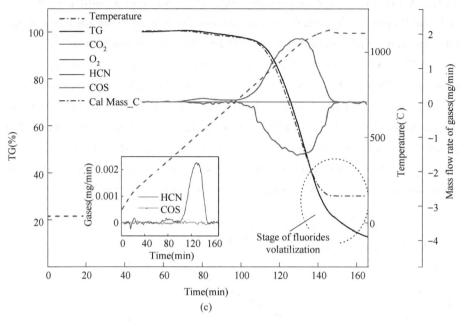

Fig. 2-105 Thermal behavior for 1 mm SCCM (sample S-0) under Ar atmosphere
(a) Mass loss, DTG, and DSC curves; (b) Mass flow rate of gases including CO_2, O_2, H_2O calculated by ECSA;
(c) Comparison between the mass loss on C element calculated by ECSA and mass loss from measurement

In conclusion, the volatilization of fluorides during treating SCCB under both Ar and Ar–O_2 atmospheres is a steady but slow process. However, the volatilization process can be radically accelerated when the carbon material in SCCB particle is burnt.

2.7.4 Thermal behavior of cyanides under Ar and combustion atmospheres

Decomposition of cyanides is a main source of producing the NO_x in the flue gas for the heat-treating SCCB process. At first, Fig. 2-106(a) shows three possible reactions of sodium cyanide at atmosphere of O_2 and the corresponding Gibbs free energy change. Thermodynamic calculation in Fig. 2-106 (a) indicates that the trend of cyanides' decomposition reaction is very sufficient, and the cyanides are more inclined to be converted into N_2, NO and NO_2 at atmosphere of O_2.

For heat-treating SCCB with particle size of 1 mm (sample S-0) under Ar–O_2 atmosphere, the variation in gases including N_2, NH_3, HCN, NO, NO_2 dertmined by TG-MS are shown in Fig. 2-106(b). By using the ECSA method, mass flow rate of each gas was calculated. Similar to the prediction by Gibbs free energy change calculation of cyanides decomposition reactions as shown in Fig. 2-106(a), the cyanides in the SCCB were mainly, under Ar–O_2

atmosphere, converted to gases of N_2, NO and NO_2. There was also a small part of HCN being produced. Proportionally, around three quarters of the cyanides were converted to the N_2, and the rest were mainly converted to the NO, according to the quantitative caculation. The amount of NO_2 and HCN produced are relatively small. All the gases released quickly at temperature ranged 750–1100 ℃ (time: 100–150 min). From the variation in total conversion rate of cyanides as shown in Fig. 2-106(c), conversion of cyanides starts actually at around 250 ℃, begins to accelerate at around 800 ℃. A nearly 100% conversion rate is obtained at 1100 ℃. Notably, a concident changing pattern between conversions of cyanides and carbon is shown in Fig. 2-106(c), which implies that cyanides has been embedded into the structure of carbon molecular network, and form a uniform distribution in whole particle.

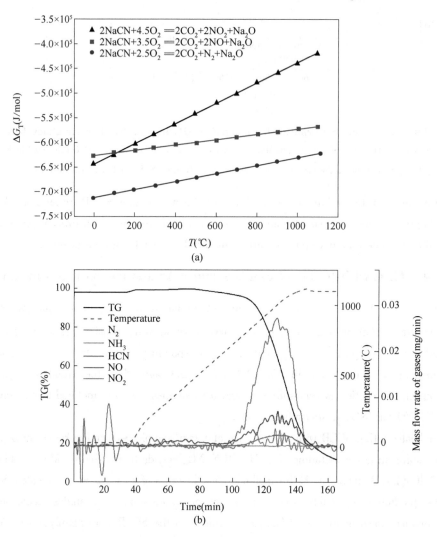

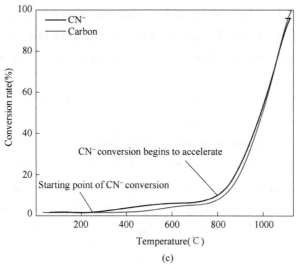

Fig. 2-106 Thermodynamics and behaviors of cyanides conversion for 1 mm SCCM (sample S-0) under Ar−O_2 atmosphere

(a) Possible decomposition reactions and the corresponding Gibbs free energy change;

(b) Variation in released gases of N_2, NH_3, HCN, NO, NO_2;

(c) Variation in total conversion rate of cyanides' decomposition and carbon combustion

Considering the structure of cyanides embedded into the carbon molecular network, particle size could affect the decompostion of cyanides during heat-treating SCCB. Thus the measurement for heat-treating SCCB with particle size of 100 μm (sample S-1) under Ar−O_2 atmospheres was carried out. Fig. 2-107(a) shows the variation in gases of N_2, NH_3, HCN, NO, NO_2 during conversion of cyanides under Ar−O_2 atmosphere. It can be seen that the variation in each gas for sample S-1 is similar to that for sample S-0, but the amount of each gas released in a unit time from sample S-1 is lower than that from sample S-0. While the decompostion of cyanides shifts towards a more lower but wide-range temperatures. All the gases released quickly at temperature ranged 350−900 °C (time:60−130 min). Quantitavely, Fig. 2-107(b) compares the variation in total conversion rate of cyanides between samples S-0 and S-1. There is also a great coincident trend between conversion of cyanides and combustion of carbon for sample S-1 as sample S-0. Since this greatly coincident relationship for both samples, it should be expected that the carbon combustion plays a significant effect on decomposition of cyanides in SCCB particle.

Generally, the combustion rate of carbon is controlled, separately or in combination, by chemical-reaction, pore-diffusion processes, or bulk-diffusion processes. The intrinsic kinetics for combustion of the used SCCB has been examined to be controlled in combination of the above three processes over a range of temperatures measured in this study. The decrease in particle size could increase the gas-solid contact area and improve the diffusion effects of O_2 into the particle, resulting into an increase in carbon combustion rate and then a quick decomposition of cyanides at lower temperature. In the other hand, actually, the temperature required for decomposition reaction of cyanides has not been intrinsically lowered, but is integratedly related to the factors such as carbon combustion, particle size, and pore structure in particle.

Lastly, the influence of atmosphere on the decomposition of cyanides is investigated. Fig. 2-107(c) shows the variation in released gases of N_2, NH_3, HCN, NO, NO_2 for heat-treating sample S-0 under Ar atmosphere, to make a comparsion with those under Ar-O_2 atmosphere. The gas composition measured under Ar atmosphere shows that only NO was produced, which was mainly resulted from the reaction of CN^- with O elements in SCCB particle.

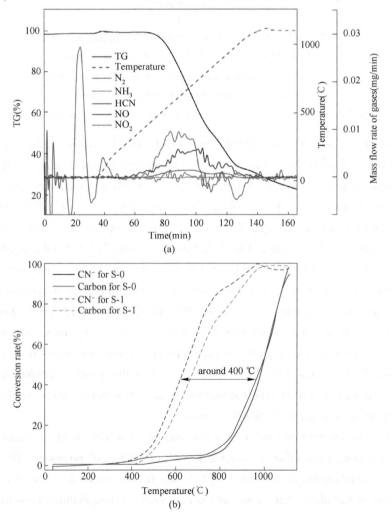

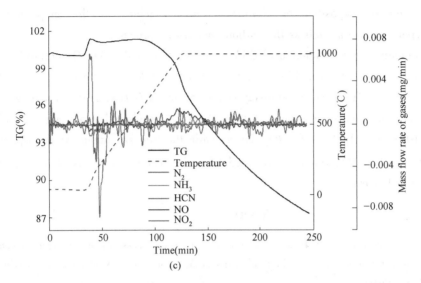

Fig. 2-107 Thermal behavior of cyanides conversion under varying conditions
(a) Variation in gases of N_2, NH_3, HCN, NO, NO_2 for heat-treating 100 μm SCCM (sample S-1) under $Ar-O_2$ atmosphere; (b) Comparison between total conversion rate of cyanides' decomposition and carbon combustion for sample S-0 and S-1; (c) Variation in gases for heat-treating sample S-0 under Ar atmosphere

In conclusion, decomposition of cyanides is greatly related to the particle size and atmosphere. The cyanides in the SCCB can be effectively decomposed to produce N_2, NO, NO_2, and HCN at the high temperature and at oxygen atmosphere. The cyanides in the SCCB with small particle size can be completely decomposed at a lower and wide-range temperatures.

The thermal behaviors of release and transformation of fluorides and cyanides during thermal treatment of SCCB were basically measured by adopting the TG/DSC-MS analysis combining a novel data-processing method of ECSA in this paper. The variation in composition and content of the flue gas under varying conditions were qualitatively and quantitatively obtained and analyzed in detail. The main conclusions are summarized as follows:

(1) During the thermal treatment, the fluorides can only be volatilized when the temperature is beyond their respective melting point and is almost not affected by the combustion of carbon in the SCCB particle.

(2) The fluorides volatilization is a steady but slow process under both Ar and $Ar-O_2$ atmospheres. And, the fluorides volatilization can be radically accelerated when the carbon material is burnt.

(3) The cyanides can be effectively decomposed at the high temperature and at oxygen

atmosphere. And, decomposition of cyanides is greatly related to the temperature, atmosphere, particle size as well as the carbon combustion process.

(4) The quantitative caculation by ECSA indicates that under $Ar-O_2$ atmosphere, around three quarters of the cyanides were converted to the N_2, and the rest was mainly converted to the NO. There was a relatively small amount of NO_2 and HCN produced.

(5) The flue gas composition measured under $Ar-O_2$ atmosphere is more complicated, mainly consisting of CO_2, N_2, NO, NO_2, HCN; while the flue gas under Ar atmosphere only consists of CO_2, NO, which are resulted from the reaction of C and CN^- with O elements in SCCB particle.

References

[1] FU Z H, SONG Z K, CHEN X H, et al. Brief analysis on energy conservation and environmental protection industry development in China[J]. Strategic Study of Chinese Academy of Engineering, 2015, 17(8):75-80.

[2] LI W X, CHEN X P. Analysis of pollution and countermeasures of aluminum electrolytic spent potlining[J]. Nonferrous Metallurgical Energy Conservation, 2008, 1:6-11.

[3] BYERS R L. Disposal of spent potlining research and development[J]. Light Metals, 1982: 1023-1030.

[4] GOMES V, DRUMOND P Z, NETO J O P, et al. Co-processing at cement plant of spent potlining from the aluminum industry[J]. Light Metals, 2005:1057-1063.

[5] RUSTAD I, KARSTENSEN K H, ØDEGRD K E. Disposal options for spent potlining[J]. Waste Management Series, 2000, 1:617-632.

[6] VENANCIO L C, SOUZA J A, MACEDO E N, et al. Residues recycling: Reduction costs and helping the environment[J]. JOM, 2010, 62(9):41-45.

[7] GAO L, MOSTAGHEL S, RAY S, et al. Using spl(spent pot-lining) as an alternative fuel in metallurgical furnaces[J]. Metallurgical & Materials Transactions E, 2016, 3(3):179-188.

[8] AGRAWAL A, SAHU K K, PANDEY B D. Solid waste management in non-ferrous industries in india[J]. Resources Conservation and Recycling, 2004, 42(2):99-120.

[9] HOLYWELL G, BREAULT R. An overview of useful methods to treat, recover, or recycle spent potlining[J]. JOM, 2013, 65(44):1441-1451.

[10] GIVENS H L. Using spent potliner as a fuel supplement in coal-fired power plants[J]. JOM, 1989, 41(3):57-59.

[11] PAWLEK R P. Water soluble components, landfill and alternative solutions[J]. Light Metals, 1993;399-405.

[12] BIRRY L, LECLERC S, POIRIER S. The LCL & L process: A sustainable solution for the treatment and recycling of spent potlining[J]. Light Metals, 2016:467-471.

[13] LI N. Recycle of spent pot-lining with low carbon grade by floatation[J]. Advanced Materials Research, 2014, 881:1660-1664.

[14] DZIKUNU P, ARTHUR E K, GIKUNOO E. Successive selective leaching procedures for valorization of spent pot lining carbon[J]. Process Safety and Environmental Protection, 2023, 169:1-12.

[15] LISBONA D F, SOMERFIELD C, STEEL K M. Leaching of spent pot-lining with aluminium nitrate and nitric acid: Effect of reaction conditions and thermodynamic modelling of solution speciation[J]. Hydrometallurgy, 2013, 134-135:132-143.

[16] SATERLAY A J, HONG Q, COMPTON R G, et al. Ultrasonically enhanced leaching: Removal and destruction of cyanide and other ions from used carbon cathodes[J]. Ultrasonics Sonochemistry, 2000, 7:1-6.

[17] LI N, GAO L, CHATTOPADHYAY K. Migration behavior of fluorides in spent potlining during vacuum distillation method[J]. Light Metals, 2019:867-872.

[18] MAZUMDER B, DEVI S R. Adsorption of oils heavy metals and dyes by recovered carbon powder from spent pot liner of aluminum smelter plant [J]. Journal of Environmental Chemical Engineering, 2008, 50:203-206.

[19] MAZUMDER B, DEVI S R. Conversion of byproduct carbon obtained from spent pot liner[J]. Journal of Applied Chemistry, 2013, 3:24-30.

[20] WOMACK R K. Using the centrifugal method for the plasma-arc vitrification of waste[J]. JOM, 1999, 51:14-16.

[21] YU D, CHATTOPADHYAY K. Numerical simulation of copper recovery from converter slags by the utilisation of spent potlining (SPL) from aluminium electrolytic cells [J]. Canadian Metallurgical Quarterly, 2016, 55:251-260.

[22] LIU F Q, GU S Q. R&D and industrial application of new structure electrolyzer technology[J]. Journal of Materials and Metallurgy, 2010, 9(1):17-19.

[23] ZHAO L, LIU F Q, MAO J H, et al. Study on mechanical wear resistance of cathode carbon block for aluminum electrolysis[J]. Light Metals, 2005, 12:25-28.

[24] ZHANG G, SUN G, LIU J Y, et al. Thermal behaviors of fluorine during (Co-) incinerations of spent potlining and red mud: Transformation, retention, leaching and thermodynamic modeling analyses[J]. Chemosphere, 2020, 249:126204.

[25] ZHANG G, SUN G, CHEN Z H, et al. Water-soluble fluorine detoxification mechanisms of spent potlining incineration in response to calcium compounds[J]. Environmental Pollution, 2020, 266:115420.

[26] ANDRADE L F, DAVIDE L C, GEDRAITE L S, et al. Genotoxicity of SPL (spent pot lining) as measured by Tradescantia bioassays[J]. Ecotoxicology and Environmental Safety, 2011, 74:2065-2069.

[27] YU D W, PAKTUNC D. Carbothermic reduction of chromite fluxed with aluminum spent potlining [J]. Transactions of Nonferrous Metals Society of China, 2019, 29:200-212.

[28] LI Y J, ZHAO J G, WANG W W, et al. Research status of electrolyte corrosion resistance of Si_3N_4 bonded SiC[J]. Light Metals, 2006, 7:25-28.

[29] GE S, YIN Y C. Study on damage mechanism of Si_3N_4 bonded SiC materials in aluminum electrolysis

cell[J]. Light Metals,2008,5:58-61.

[30] WANG Z Y,XIAO Y M,ZHANG F B. Application test of dry anti-seepage material on aluminum electrolysis cell[J]. Light Metals,1999,6:33-36.

[31] COURBARIAUX Y, CHAOUKI J, GUY C. Update on spent potliners treatments: Kinetics of cyanides destruction at high temperature[J]. Industrial & Engineering Chemistry Reseacrch, 2004,43:5828-5837.

[32] SLEAP S B, TURNER B D, SLOAN S W. Kinetics of fluoride removal from spent pot liner eachate(SPLL) contaminated groundwater[J]. Journal of Environmental Chemical Engineering, 2015,3:2580-2587.

[33] TSCHÖPE K,SCHØNING C, RUTLIN J, et al. Chemical degradation of cathode linings in hall-héroult cells—An autopsy study of three spent pot linings[J]. Metallurgical and Materials Transactions B,2011,43:290-301.

[34] PALMIERI M J,ANDRADE-VIEIRA L F,TRENTO M V C,et al. Cytogenotoxic effects of spent pot liner(SPL) and its main components on human leukocytes and meristematic cells of Allium cepa[J]. Water Air and Soil Pollution,2016,227:1-10.

[35] LISBONA D F, STEEL K. Recovery of fluoride values from spent pot-lining: Precipitation of an aluminium hydroxyfluoride hydrate product[J]. Separation and Purification Technology, 2018, 61:182-192.

[36] SHI Z, LI W, HU X, et al. Recovery of carbon and cryolite from spent pot lining of aluminium reduction cells by chemical leaching[J]. Transactions of Nonferrous Metals Society of China, 2012,22:222-227.

[37] PONG T K, ADRIEN R J, BESIDA J, et al. Spent potlining—A hazardous waste made safe[J]. Process Safety and Environmental Protection,2000,78:204-208.

[38] BROOKS D G,CUTSHALL E R,BANKER D B,et al. Thermal treatment of spent potlining in a kiln[J]. Light Metal,1992,283:1044-1048.

[39] LI W X, CHEN X P. Development status of processing technology for spent potlining in China [J]. Light Metal,2010,859:1064-1066.

[40] SILVEIRA B I, DANTAS A E, BLASQUEZ J E, et al. Characterization of inorganic fraction of spent potliners: Evaluation of the cyanides and fluorides content[J]. Journal of Hazardous Materials,2002,89:177-183.

[41] SILJAN O J, SCHÖNING C, GRANDE T. State-of-the-art alumino-silicate refractories for Al electrolysis cells[J]. JOM,2002,54:46-55.

[42] TABEREAUX A T, BROWN J H, ELDRIDGE I J, et al. Erosion of cathode blocks in 180 kA prebake cells[J]. Light Metal,1999:999-1004.

[43] BREAULT R,POIRIER S,HAMEL G, et al. A "green" way to deal with spent pot lining[J]. Aluminium Int. Today J. Aluminium Prod. Process,2011,23:22-24.

[44] HUANG S. The treatment of SPL and its technical analysis[J]. Light Met. ,2009:29-34.

[45] ANDRADE L F, DAVIDE L C, GEDRAITE L S. The effect of cyanide compounds, fluorides, aluminum, and inorganic oxides present in spent pot liner on germination and root tip cells of

Lactuca sativa[J]. Ecotoxicol. Environ. Saf. ,2010,73:626-631.

[46] TURNER B D, BINNING P J, SLOAN S W. A calcite permeable reactive barrier for the remediation of fluoride from spent potliner (SPL) contaminated groundwater [J]. J. Contam. Hydrol. ,2008,95:110-120.

[47] FLORES I V, FRAIZ F, LOPES JUNIOR R A, et al. Evaluation of spent pot lining (SPL) as an alternative carbonaceous material in ironmaking processes[J]. J. Mater. Res. Technol. ,2019,8: 33-40.

[48] YU D W, CHATTOPADHYAY K. Enhancement of the nickel converter slag-cleaning operation with the addition of spent potlining[J]. Int. J. Min. Met. Mater. ,2018,25:881-891.

[49] XU Y T, YANG B, LIU X M, et al. Investigation of the medium calcium based non-burnt brick made by red mud and fly ash: Durability and hydration characteristics [J]. Int. J. Min. Met. Mater. ,2018,26:983-991.

[50] ROBSHAW T J, ATKINSON D, HOWSE J, et al. Recycling graphite from waste aluminium smelter Spent Pot Lining into lithium-ion battery electrode feedstock [J]. Cleaner Production Letters,2022,2:100004.

[51] XIE M Z, LI R B, ZHAO H L, et al. Detoxification of spent cathode carbon blocks from aluminum smelters by joint controlling temperature-vacuum process[J]. J. Clean. Prod. ,2020,249:1-10.

[52] SAVOV L, JANKE D. Evaporation of Cu and Sn from induction stirred iron-based melts treated at reduced pressure[J]. ISIJ Int. ,2000,40:95-104.

[53] NASR M I, OMAR A A, KHEDR M H, et al. Effect of nickel oxide doping on the kinetics and mechanism of iron oxide reduction[J]. ISIJ Int. ,1995,35:1043-1049.

[54] CAO X Z, SHI Y Y, ZHAO S, et al. Recovery of valuable components from spent pot-lining of aluminium electrolytic reduction cells [J]. J. Northeast Univ. Nat. Sci. , 2014, 35 (12): 1746-1749.

[55] BAO L F, ZHAO J X, TANG W D, et al. Separation and recycling use of waste cathode in aluminium electrolysis cells[J]. China Nonferrous Metall. ,2014,43(3):51-54.

[56] MA J L, SHANG X F, MA Y P, et al. Directions for development of hazardous waste treatment technologies in primary aluminum industry[J]. Environ. Prot. Chem. Ind. ,2016,36(1):11-15.

[57] HITTNER H J, BYERS L R, LEES J N, et al. Rotary kilntreatment of potliner: US Patent, 5711018[P]. 1998-01-20.

[58] BARRILLON E, PERSONNET P, BONTRON J. Process for the thermal shock treatment of spent pot linings obtained from hall-heroult electrolytic cells: US Patent, 5245115[P]. 1993-09-14.

[59] GROLMAN R J, HOLYWELL G C, KIMMERLE F M, et al. Recycling of spent pot linings: US Patent, 5740559[P]. 1995-11-28.

[60] CHEN X P, LI W X, ZHOU J M, et al. Studying on the toxicity of spent potline in aluminum electrolysis[J]. Light Met, 2005(12):33-36.

[61] ZHAI X J, QIU Z X. Applying flotation to separate electrolyte from spent carbon of aluminum electrolysis[J]. Nonferrous Met, 1993,45(2):38-42.

[62] BELL N, ANDERSEN J N, LAM H K H. Process for the utilization of waste materials from

primary aluminum reduction systems: US Patent, 4113832[P]. 1978-12-12.

[63] RENÓ M L G, TORRES F M, da SILVA R J, et al. Exergy analyses in cement production applying waste fuel and mineralizer[J]. Energy Convers. Manage, 2013, 75: 98-104.

[64] GHENAI C, INAYAT A, SHANABLEH A, et al. Combustion and emissions analysis of Spent Pot Lining(SPL) as alternative fuel in cement industry[J]. Sci. Total. Environ. , 2019, 684: 519-526.

[65] ZHANG B, QI H, RUAN L M. Two-dimensional simulation for the effective thermal conductivity of heat-sealing porous material[J]. J. Eng. Thermophys, 2012, 33(7): 1229-1235.

[66] GORAI B, JANA R K, PREMCHAND K. Characteristics and utilisation of copper slag—A review [J]. Resour. Conserv. Recycl. , 2003, 39(4): 299-313.

[67] SINGH J, SINGH S P. Geopolymerization of solid waste of non-ferrous metallurgy—A review[J]. J. Environ. Manage. , 2019, 251: 109571.

[68] DANDAUTIYA R, SINGH A P. Utilization potential of fly ash and copper tailings in concrete as partial replacement of cement along with life cycle assessment[J]. Waste Manag, 2019, 99: 90-101.

[69] NUNEZ P. Developing guidance to support sustainable spent pot lining(SPL) management across the aluminum industry[J]. JOM, 2020, 72(10): 1-7.

[70] MUGFORD C, GIBBS J L, BOYLSTEIN R. Elemental properties of copper slag and measured airborne exposures at a copper slag processing facility[J]. J. Occup. Environ. Hyg. , 2017, 14 (8): 120-126.

[71] ALTUNDOĞAN H S, TÜMEN F. Metal recovery from copper converter slag by roasting with ferric sulphate[J]. Hydrometallurgy, 1997, 44(1): 261-267.

[72] BELLEMANS I, DE WILDE E, MOELANS N, et al. Metal losses in pyrometallurgical operations—A review[J]. Adv. Colloid Interface Sci. , 2018, 255: 47-63.

[73] BROOKS C S. Metal recovery from industrial wastes[J]. JOM, 1986, 38(7): 50-57.

[74] KUMAR A, SAMADDER S R, ELUMALAI S P. Recovery of trace and heavy metals from coal combustion residues for reuse and safe disposal: A review[J]. JOM, 2016, 69(9): 2413-2417.

[75] ALLOWAY B J. Heavy Metals in Soils[M]. Springer, Dordrecht, 2013.

[76] KHALID M K, HAMUYUNI J, AGARWAL V, et al. Sulfuric acid leaching for capturing value from copper rich converter slag[J]. J. Clean. Prod. , 2019, 215(1): 1005-1013.

[77] ZHU Z, ZHANG W, PRANOLO Y, et al. Separation and recovery of copper, nickel, cobalt and zinc in chloride solutions by synergistic solvent extraction[J]. Hydrometallurgy, 2012, 127/128: 1-7.

[78] SHEN H T, FORSSBERG E. An overview of recovery of metals from slags[J]. Waste Manag, 2003, 23: 933-949.

[79] TIJSSELING L T, DEHAINE Q, ROLLINSON G K H J. Flotation of mixed oxide sulphide copper-cobalt minerals using xanthate, dithiophosphate, thiocarbamate and blended collectors[J]. Miner. Eng. , 2019, 138: 246-256.

[80] LEE K, ARCHIBALD D, MCLEAN J, et al. Flotation of mixed copper oxide and sulphide minerals with xanthate and hydroxamate collectors[J]. Miner. Eng. , 2009, 22: 395-401.

[81] YIN F, XING P, LI Q, et al. Magnetic separation-sulphuric acid leaching of Cu-Co-Fe matte obtained from copper converter slag for recovering Cu and Co[J]. Hydrometallurgy, 2014, 149: 189-194.

[82] YANG X, ZHANG J L, ZHANG J K, et al. Efficient recovery of copper and cobalt from the matte-slag mixture of ISA furnace by injection of coke and pyrite[J]. Metall. Mater. Trans. B, 2018, 49(6):3118-3126.

[83] LI Y, CHEN Y M, TANG C B, et al. Co-treatment of waste smelting slags and gypsum wastes via reductive-sulfurizing smelting for valuable metals recovery[J]. J. Hazard. Mater., 2017, 322(15):402-412.

[84] XIE W M, ZHOU F P, LIU J Y, et al. Synergistic reutilization of red mud and spent pot lining for recovering valuable components and stabilizing harmful element[J]. J. Clean. Prod., 2020, 243:118624.

[85] SOMERVILLE M, DAVIDSON R, WRIGHT S, et al. Liquidus- and primary-phase determinations of slags used in the processing of spent pot lining[J]. J. Sustain. Metall., 2016, 3(3):486-494.

[86] LIU F Q, XIE M Z, LIU W, et al. Footprint of harmful substances in spent pot lining of aluminum reduction cell[J]. Trans. Nonferrous Met. Soc. China, 2020, 30(7):1956-1963.

[87] JAWI M A, CHOW C M, PUJARI S, et al. Environmental benefits of using spent pot lining (SPL) in cement production[J]. Light Metals, 2020:1251-1260.

[88] PALMIERI M J, LUBER J, ANDRADE-VIEIRA L F, et al. Cytotoxic and phytotoxic effects of the main chemical components of spent pot-liner: A comparative approach[J]. Mutat. Res Gen. Tox. En., 2014, 763(15):30-35.

[89] LI N, JIANG Y, LV X, et al. Vacuum distillation-treated spent potlining as an alternative fuel for metallurgical furnaces[J]. JOM, 2019, 71(9):2978-2985.

[90] XIE M Z, GUO X Y, LIU W, et al. Phase transition of waste silicon carbide side block from aluminum smelters during vacuum high-temperature detoxification process[J]. JOM, 2020, 72(7):2697-2704.

[91] DU K, LI H X, ZHANG M M. Calculation of distribution coefficients of cobalt and copper in Matte and slag phases in reduction-vulcanization process of copper converter slag[J]. JOM, 2017, 69(11):1-4.

[92] ZHAO H L, LIU F Q, XIE M Z, et al. Recycling and utilization of spent potlining by different high temperature treatments[J]. J. Clean. Prod., 2021, 289:125704.

[93] HONG S, LIU W, LIU F Q. Preliminary study on reduction and extraction of copper and cobalt from copper converter slag by using waste cathode carbon block[J]. Light Metals, 2019, 8: 41-45.

[94] ROCHA V C D, SILVA M L D, BIELEFELDT W V, et al. Assessment of viscosity calculation for calcium-silicate based slags using computational thermodynamics[J]. Rem Int. Eng. J., 2018, 71(2):243-252.

[95] PELTON A D, CHARTRAND P. The modified quasi-chemical model: Part II. Multicomponent solutions[J]. Metall. Mater. Trans. A, 2001, 32(6):1355-1360.

[96] NEMCHINOVA N V, BARANOV A N, BARAUSKAS A E. Choosing the reagent to leach fluorine from spent pot lining of aluminum electrolysis cells[J]. Materials Science Forum, 2022, 1052: 488-492.

[97] THOMAS J, GLUSKOTER H J. Determination of fluoride in coal with the fluoride ion-selective electrode[J]. Anal. Chem. , 1974, 46(9): 1831-1832.

[98] SRIVASTAVA R K, JOZEWICZ W. Flue gas desulfurization: The state of the art[J]. J. Air Waste Manage. , 2001, 51(12): 1676-1688.

[99] ISLAM M, PATEL P K. Evaluation of removal efficiency of fluoride from aqueous solution using quick lime[J]. J. Hazard. Mater. , 2007, 143(1/2): 303-310.

[100] KANG J H, GOU X Q, HU Y H, et al. Efficient utilisation of flue gas desulfurization gypsum as a potential material for fluoride removal[J]. Sci. Total Environ. , 2019, 649: 344-352.

[101] XIE M Z, LV H, LU T T, et al. Characteristic analysis of hazardous waste from aluminum reduction industry[J]. Light Met. , 2020: 1261-1266.

[102] MAHINROOSTA M, ALLAHVERDI A. Hazardous aluminum dross characterization and recycling strategies: A critical review[J]. J. Environ. Manage. , 2018, 223: 452-468.

[103] LIAO C Z, SU M, MA S, et al. Immobilization of lead in cathode ray tube funnel glass with beneficial use of red mud for potential application in ceramic industry[J]. ACS Sustain. Chem. Eng. , 2018, 6(11): 14213-14220.

[104] ISHAK R, LAROCHE G, LAMONIER J F, et al. Characterization of carbon anode protected by low boron level: An attempt to understand carbon-boron inhibitor mechanism[J]. ACS Sustain. Chem. Eng. , 2017, 5(8): 6700-6706.

[105] SUN G, ZHANG G, LIU J Y, et al. Thermogravimetric and mass-spectrometric analyses of combustion of spent potlining under N_2/O_2 and CO_2/O_2 atmospheres[J]. Waste Manag. , 2019, 87: 237-249.

[106] MANN V, PINGIN V, ZHERDEV A, et al. SPL recycling and re-processing[J]. Light Met. , 2017: 571-578.

[107] LISBONA D F, SOMERFIELD C, STEEL K M. Leaching of spent pot-lining with aluminum anodizing wastewaters: Fluoride extraction and thermodynamic modeling of aqueous speciation [J]. Ind. Eng. Chem. Res. , 2012, 51(25): 8366-8377.

[108] XIAO J, ZHANG L Y, YUAN J, et al. Co-utilization of spent pot-lining and coal gangue by hydrothermal acid-leaching method to prepare silicon carbide powder[J]. J. Clean. Prod. , 2018, 204: 848-860.

[109] SUN G, ZHANG G, LIU J Y, et al. (Co-) combustion behaviors and products of spent potlining and textile dyeing sludge[J]. J. Clean. Prod. , 2019, 224: 384-395.

[110] CIESLIK B M, NAMIESNIK J, KONIECZKA P. Review of sewage sludge management: Standards, regulations and analytical methods[J]. J. Clean. Prod. , 2015, 90: 1-15.

[111] DECASA G, MANGIALARDI T, PAOLINI A E, et al. Physical-mechanical and environmental properties of sintered municipal incinerator fly ash[J]. Waste Manage. , 2007, 27: 238-247.

[112] WANG Y W, PENG J P, DI Y Z. Separation and recycling of spent carbon cathode blocks in the

aluminum industry by the vacuum distillation process[J]. JOM,2018,70(9):1877-1882.

[113] AL-MAQBALII A,FEROZ S,RAM G,et al. Feasibility study on spent pot lining(SPL)as raw material in cement manufacture process[J]. Int. J. Environ. Chem. ,2016,2(2):18-26.

[114] KONDRATIEV V V,RZHECHITSKIY E P,ERSHOV V A,et al. Results of carrying out of researches with revealing of technological parameters of processes of recycling and neutralization of the first and second cut of the spent pot lining of electrolyzers for reception of aluminum fluoride[J]. Int. J. Appl. Eng. Res. ,2017,12(22):12801-12808.

[115] LU T T,WANG J Q,LI R B,et al. Numerical investigation on effective thermal conductivity and heat transfer characteristics in a furnace for treating spent cathode carbon blocks[J]. JOM, 2020,72(5):1971-1978.

[116] RAI A,PRABAKAR J,RAJU C B,et al. Metallurgical slag as a component in blended cement [J]. Constr. Build. Mater. ,2020,16(8):489-494.

[117] LI R B,LU T T,XIE M Z,et al. Analysis on thermal behavior of fluorides and cyanides for heat-treating spent cathode carbon blocks from aluminum smelters by TG/DSC-MS & ECSA? [J]. Ecotoxicol. Environ. Saf. ,2020,189:110015.

[118] KACIMI L,SIMON-MASSERON A,GHOMARI A,et al. Influence of NaF,KF and CaF_2 addition on the clinker burning temperature and its properties[J]. Comptes. Rendus. Chimie. ,2006,9 (1):154-163.

[119] TSCHÖPE K,SCHØNING C,GRANDE T. Autopsies of spent potlinings—A revised view[J]. Light Met. ,2009:1085-1090.

[120] LI W,CHEN X. Chemical stability of fluorides related to spent potlining[J]. Light Met. ,2008, 855-858.

Chapter 3 Anode carbon residue/electrolyte disposal technology

Improving the ecological environment, increasing resource utilization, promoting harmony between humans and nature, and implementing sustainable development are the strategies for China's economic development. In recent years, China's aluminum industry has developed rapidly. In 2019, China's primary aluminum production reached 35.93 Mt, accounting for approximately 56.7% of the world's primary aluminum production. Due to the increasing production capacity of primary aluminum in China, the waste aluminum electrolytes generated by the primary aluminum industry can be used to increase production capacity. Therefore, the environmental harm of waste aluminum electrolytes is not prominent. However, in recent years, with the control of primary aluminum production capacity in China, the new primary aluminum production capacity has decreased, resulting in a large amount of waste aluminum electrolytes being piled up without any place to use. The production of one ton of aluminum theoretically produces approximately 17 kg of fluoride salt waste aluminum electrolyte, mainly composed of cryolite. If the reasonable recycling and utilization of waste aluminum electrolytes cannot be achieved, it will not only cause waste of resources, but also cause huge harm to the environment. In recent years, the environmental pressure for human survival has been increasing, and resources are becoming increasingly scarce. The harmless treatment and comprehensive recycling of waste aluminum electrolytes are an important way to solve environmental problems in the primary aluminum industry and fully utilize resources.

At present, there is an acid leaching method for recycling waste aluminum electrolytes. However, acid leaching causes serious corrosion to production equipment, and the recovered substances are relatively single, which can easily cause secondary pollution to the environment. Moreover, most research is still in the laboratory stage, lacking industrial application. Therefore, achieving the comprehensive recycling and utilization of waste aluminum electrolyte resources will reduce the impact on the ecological environment, and is an effective way to address resource, energy, and environmental constraints in economic and social development, ensuring the circular development of aluminum industry resources.

3.1 Composition and phase

The carbon dust is generated from the anode carbon block and dropped into the reduction cell during the electrolysis process due to anode quality, electrolysis operation and electrolyte properties. The carbon dust is wrapped or floated on the surface of the electrolyte, which seriously affects the production. It is necessary to regularly remove the carbon dust from the reduction cell. Carbon dusts are generally stored as waste in the dumping site. Typical carbon dusts are shown in Fig. 3-1.

Fig. 3-1 Typical carbon dusts from the aluminum reduction industry

The phase and chemical composition of the carbon dusts were analyzed. As a result (Table 3-1), the carbon dusts contained a large amount of a fluoride salt (about 60% - 70%), alumina, and carbon.

Disposal of carbon dusts as waste will cause serious pollution to the environment[1]. Most of the fluoride salts in the carbon dusts are electrolytes that are taken together when the carbon dusts is taken. These fluoride salts are easier to separate from the carbon. There is also a small amount of fluoride salt infiltrated into the pores of the carbon dusts, which needs to be ground to separate.

Table 3-1 Chemical element composition of carbon dusts

Element	F	Al	Na	Ca	Fe	Si	Mg	C
Content(wt%)	32.26	12.91	16.34	1.08	0.52	1.70	0.82	19.68

3.2 Experiment on calcification and roasting of waste aluminum electrolyte

3.2.1 Effect of calcium oxide addition on aluminum and sodium recovery rates

The recovery and utilization of multiple elements in aluminum electrolyte of extra 0.6 Mt produced in Chinese smelting industry every year is studied. Based on chemical analysis and mineral tests for the aluminum electrolyte the influences of available lime additions on calcified roasting reactions of aluminum electrolyte are investigated and the leaching and reaction products of the calcified roasted products in caustic solution are revealed by the tests. The test results show that the multiple valuable elements in aluminum electrolyte such as sodium, aluminum, fluorine and calcium can be recovered and resource utilized by calcified roasting under the conditions of 40% – 50% of available lime and following caustic leaching for the roasted products.

With the rapid development of the national economy, the supply of resources and energy is becoming increasingly tight, and energy conservation and emission reduction have become the theme of economic development in the 21st century[2]. Improving the ecological environment, increasing resource utilization, promoting harmony between humans and nature, and implementing sustainable development are the strategies for China's economic development. In the past two decades, China's aluminum smelting industry has developed rapidly. In 2019, China's primary aluminum production reached 35.8 Mt, accounting for approximately 56.7% of the world's primary aluminum production. In the production of aluminum electrolysis, the added alumina raw material often contains about 0.35% sodium oxide, which accumulates in the aluminum electrolyte, leading to an increase in the molecular ratio and temperature of the aluminum electrolyte. To control the molecular ratio and temperature, it is necessary to add aluminum fluoride to adjust the composition of the electrolyte. Producing one ton of raw aluminum requires approximately 20 kg of aluminum fluoride. The regular addition of aluminum fluoride to the electrolytic cell can stabilize the molecular ratio of the electrolyte between 2.25 and 2.45. Low molecular ratio is beneficial for reducing the electrolysis temperature and improving current efficiency. During this process, sodium oxide reacts with aluminum fluoride to form cryolite as follows:

$$3Na_2O + 2AlF_3 \Longrightarrow 6NaF + Al_2O_3 \tag{3-1}$$

$$3NaF + AlF_3 \Longrightarrow Na_3AlF_6 \tag{3-2}$$

With the addition and consumption of alumina, long-term accumulation can lead to an increase in electrolytes in the electrolytic cell. In order to maintain a suitable electrolyte level and maintain the balance of electrolytes in the aluminum electrolytic cell, it is necessary to regularly remove electrolytes, which are in a large amount of idle and stacked state for a long time[3]. The production of one ton of aluminum theoretically generates approximately 17 kg of excess aluminum electrolytes, mainly composed of cryolite fluoride salts. China will produce approximately 600 kt of excess aluminum electrolytes annually.

At present, there are reports involving the recovery and treatment process of high lithium aluminum electrolytes, but they only focus on the recovery and utilization of lithium, without considering and paying attention to the recovery of other valuable elements in aluminum electrolytes, resulting in resource waste. Due to the use of acid leaching process to treat aluminum electrolyte, it will cause serious corrosion to production equipment, and the recovered substances are relatively single, which is prone to secondary pollution to the environment. Moreover, most research is still in the laboratory stage[4-5]. The safe treatment and resource utilization of discarded aluminum electrolysis waste is a necessary way for the green and healthy development of China's aluminum industry in the future. This can achieve efficient utilization of resources while protecting the environment.

This article uses aluminum electrolyte as raw material and calcium carbonate as a calcium source additive. Through the collaborative treatment method of "roasting activation phase inversion alkali deep leaching", it achieves the treatment and reuse of solid waste in the aluminum electrolysis industry and the recovery of valuable metal elements in the aluminum electrolyte. The process technology is simple, the energy consumption cost is low, the product quality is high, and there is no wastewater, exhaust gas, or waste residue generated during the entire treatment process. It is a green and efficient treatment method for the high value resource utilization of waste aluminum electrolyte, which has greater significance in the context of resource circulation and low-carbon economy.

The raw material used in the experiment is aluminum electrolyte from an aluminum factory in Qingtongxia, Ningxia.

Experimental reagents: Calcium carbonate (AR), sodium hydroxide (AR). Polyvinyl alcohol solution (10 g/L), sodium salicylate solution (100 g/L), barium chloride solution (30 g/L), green light phenolphthalein mixed indicator, sodium acetate acetic acid buffer solution (pH = 5.2–5.7), phenolphthalein indicator, xylenol orange indicator (3 g/L) EDTA standard solution (0.098 mol/L), zinc nitrate standard solution (0.01962 mol/L), hydrochloric acid standard solution (0.3226 mol/L), and sodium hydroxide standard solution (0.3226 mol/L).

Experimental equipment: Muffle furnace, rapid compression sample preparation crusher, constant temperature water bath, electromagnetic stirrer, vacuum filter, electronic balance, X-ray diffractometer, X-ray fluorescence spectrometer.

X-ray diffraction analysis was conducted on the experimental raw materials, and the phase analysis results are shown in Fig. 3-2. It can be seen that the main phase components of the aluminum electrolyte are Na_3AlF_6, CaF_2 and AlF_3. And X-ray fluorescence analysis was conducted on the experimental materials, and the main chemical components are shown in Table 3-2. It can be seen that the main elements of the aluminum electrolyte are F, Al, Na, etc.

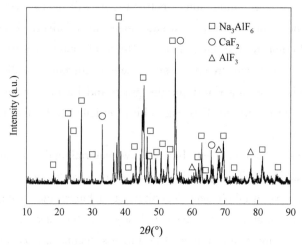

Fig. 3-2 The XRD pattern of aluminum electrolyte

Table 3-2 Composition and content of chemical elements in aluminum electrolyte

(%)

Elements	F	Na	Al	Ca	Others	Total
Content	54.10	24.89	13.93	1.9	5.18	100

The aluminum electrolyte raw materials used in the experiment were mechanically crushed, screened, and ground with calcium carbonate on a planetary ball mill at a speed of 100 r/min for 2 h. Then, the mixed materials were made into columnar samples at 40 MPa, and the samples were placed in a muffle furnace for high-temperature sintering at different temperatures for 1 h. Titrate and measure the alumina and caustic soda in the leaching solution, and calculate the recovery rates of aluminum and sodium.

The aluminum and sodium contents in the filtrate are determined according to the chemical analysis method of sodium aluminate solution. Transfer 5 mL of the solution to be tested into a 100 mL volumetric flask, dilute with water to the mark, and mix well. This solution is used for the determination of alkali, total alkali, and alumina content.

Divide 10 mL of the above solution into a 500 mL conical flask, add 60 mL of barium chloride mixed solution and 10 mL of sodium salicylate solution, mix well, add 20 drops of green light phenolphthalein mixed indicator, and immediately titrate with hydrochloric acid standard solution until the end point is grayish green. Attention: During titration, sufficient vibration should be used to avoid amorphous aluminum hydroxide adsorbing sodium hydroxide and indicator. For solutions with high carbon alkali content, the addition of barium chloride mixture should be 100 mL. Divide 10 mL of the above solution into a 500 mL conical flask filled with an appropriate amount of EDTA standard solution beforehand, add 15 mL of hydrochloric acid standard solution, wash the bottle wall with water, dilute to about 100 mL, heat and boil for 3 min, remove, add 6 drops of phenolphthalein indicator, and titrate with sodium hydroxide standard solution until it turns slightly red. Immediately add 20 mL of sodium acetate acetic acid buffer solution, cool to room temperature with running water, add 4-5 drops of xylenol orange indicator, and titrate with zinc nitrate standard solution until the solution turns rose red as the endpoint.

The effect of effective calcium oxide addition on the composition of the calcined product and the analysis results are shown in Fig. 3-3. It can be seen that when the effective calcium oxide addition is 10%(wt), the main phases of the product are CaF_2, NaF, $CaO \cdot 6Al_2O_3$. Calcium oxide reacts with cryolite to generate alumina, which then combines with the generated alumina to form $CaO \cdot 6Al_2O_3$. The reaction equation is shown in Eq. (3-6). At this point, there is no alumina in the phase, indicating that when the effective amount of calcium oxide is 10%(wt), both cryolite and aluminum fluoride in the electrolyte have reacted with calcium oxide, and the aluminum element exists in the form of alumina in calcium aluminate. When the effective amount of calcium oxide is increased to 30%(wt), the main components of the calcined clinker change from CaF_2, NaF, $CaO \cdot 6Al_2O_3$ to CaF_2, NaF, $CaO \cdot 2Al_2O_3$(as shown in reaction Eq. (3-7)).

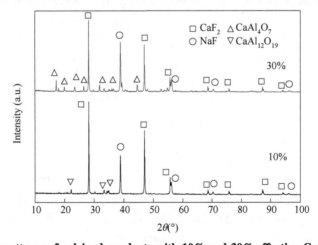

Fig. 3-3　XRD patterns of calcined products with 10% and 30% effective CaO contents(wt)

From Fig. 3-4, it can be seen that when the effective amount of calcium oxide added to the mixture reaches 40%(wt), the main components of the calcined clinker change from CaF_2, NaF, $CaO \cdot 2Al_2O_3$ to CaF_2, NaF, $12CaO \cdot 7Al_2O_3$. Calcium aluminate exists in the form of $12CaO \cdot 7Al_2O_3$.

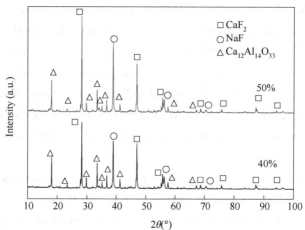

Fig. 3-4　XRD patterns of calcined products with 40% and 50% effective CaO contents(wt)

As shown in Fig. 3-5, when the effective amount of calcium oxide reaches 60%(wt), $Ca(OH)_2$ phase appears in the calcined clinker. Due to the decomposition of $Ca(OH)_2$ at 500–600 ℃, the $Ca(OH)_2$ phase present in the calcined clinker may be formed by the reaction of the calcium oxide phase in the sample with water in the air before detection. When the effective amount of calcium oxide added is 60%(wt), the amount of calcium oxide added is excessive and will have an impact on the subsequent alkaline solution dissolution of alumina experiment, resulting in a decrease in its dissolution rate.

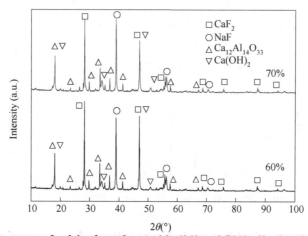

Fig. 3-5　XRD patterns of calcined products with 60% and 70% effective CaO contents(wt)

The main component of aluminum electrolyte is cryolite, in addition to a small amount of alumina, calcium fluoride, and excess aluminum fluoride. Due to the fact that aluminum in aluminum electrolytes mainly exists in the form of cryolite, the crystal structure of cryolite is composed of an isolated octahedral AlF_6^{3-} which combines with Na^+ at six vertices to form a regular octahedron with good stability and is generally difficult to decompose. When the aluminum electrolyte is mixed with calcium source additives for roasting[6-7], the highly active calcium oxide produced by the decomposition of calcium carbonate will react with cryolite to generate alumina (as shown in reaction Eq. (2-3) and Eq. (2-4)). The generated alumina continues to combine with calcium oxide to form calcium aluminate (reaction Eq. (3-5)), and the alumina in calcium aluminate can be dissolved through alkaline solution.

Table 3-3 shows the chemical reactions that may occur during the roasting experiment and their thermodynamic calculations (1000 ℃). The standard Gibbs free energy of each reaction in the table is far less than zero, which means that all conditions for occurrence are met. It can be seen that using calcium carbonate as a calcium source additive to treat and recycle aluminum electrolytes is thermodynamically feasible.

Table 3-3 Main reaction process and its standard Gibbs free energy change

Reaction equation	ΔG (kJ/mol)	No.
$2Na_3AlF_6 + 3CaO = Al_2O_3 + 6NaF + 3CaF_2$	−53.43	Eq. (3-3)
$3CaO + 2AlF_3 = Al_2O_3 + 3CaF_2$	−451.46	Eq. (3-4)
$12CaO + 7Al_2O_3 = 12CaO \cdot 7Al_2O_3$	−308.21	Eq. (3-5)
$CaO + 6Al_2O_3 = CaO \cdot 6Al_2O_3$	−67.23	Eq. (3-6)
$CaO + 2Al_2O_3 = CaO \cdot 2Al_2O_3$	−49.19	Eq. (3-7)

Under the same conditions (calcination temperature of 1000 ℃, calcination time of 60 min), the effect of different effective calcium oxide additions on the phase composition of the leaching product is shown in Fig. 3-6 and Fig. 3-7. When the effective amount of calcium oxide added is less than 60% (wt), the phase of the leaching product is relatively single as CaF_2; When the amount of effective calcium oxide added to the material system continues to increase to 60% (wt), the leaching product begins to exhibit $Ca(OH)_2$ phase, which affects the purity of the final recovered calcium fluoride.

During the dissolution experiment of alumina, calcium aluminate in the calcined clinker will react with the alkaline solution to form sodium aluminate, which enters the alkaline solution. At the same time, the generated $Ca(OH)_2$ will react with the F^- in the solution to form CaF_2 precipitation (the reaction equations are shown in Eq. (3-8) −

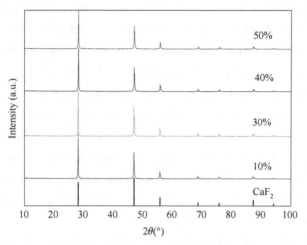

Fig. 3-6 XRD patterns of leaching products with 10%–50% effective CaO contents (wt)

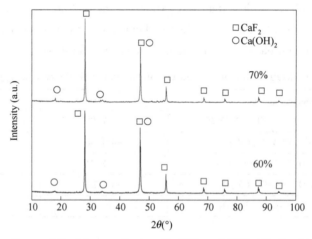

Fig. 3-7 XRD patterns of leaching products with 60% and 70% effective CaO contents (wt)

Eq. (3-11)).

$$12CaO \cdot 7Al_2O_3 + 14NaOH + 33H_2O = 14NaAl(OH)_4 + 12Ca(OH)_2 \quad (3\text{-}8)$$
$$Ca(OH)_2 + 2NaF = CaF_2 + 2NaOH \quad (3\text{-}9)$$
$$CaO \cdot 6Al_2O_3 + 12NaOH + 19H_2O = 12NaAl(OH)_4 + Ca(OH)_2 \quad (3\text{-}10)$$
$$CaO \cdot 2Al_2O_3 + 12NaOH + 7H_2O = 4NaAl(OH)_4 + Ca(OH)_2 \quad (3\text{-}11)$$

The effect of effective calcium oxide addition on the recovery rate of aluminum sodium is shown in Fig. 3-8 under the conditions of dissolution temperature of 90 ℃, dissolution time of 90 min, NaOH concentration of 5%, and liquid-solid ratio of 10∶1. As shown in Fig. 3-8, with the increase of effective calcium oxide addition, the recovery of aluminum and sodium first increases and then decreases. This is because in aluminum calcium

3.2 Experiment on calcification and roasting of waste aluminum electrolyte

compounds, the solubility of $12CaO \cdot 7Al_2O_3$ in alkaline solution is better than that of $CaO \cdot 2Al_2O_3$, and the solubility of $CaO \cdot 2Al_2O_3$ is also better than that of $CaO \cdot 6Al_2O_3$[8]. $12CaO \cdot 7Al_2O_3$ is composed of densely packed sub nanometer sized cages, and the spatial distance between O^{2-} and Ca^{2+} in $12CaO \cdot 7Al_2O_3$ is 1.5 times larger than the sum of the radii of these two ions (0.24 nm), resulting in the formation of a pore like structure with a diameter of 0.1 nm, through which ion groups can enter and exit the "cage"[9-10]. During the dissolution experiment, the alkaline solution can enter and exit through this micropore, thereby improving the Al_2O_3 leaching rate of $12CaO \cdot 7Al_2O_3$.

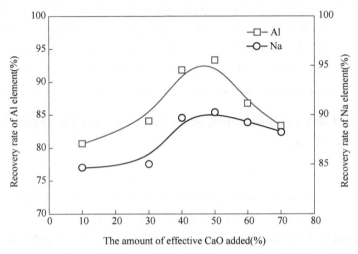

Fig. 3-8 Effect of effective CaO addition on recovery of Al and Na

When the effective amount of calcium oxide is above 60% (wt), the excess CaO in the calcined clinker will inhibit the dissolution reaction of calcium aluminate in the alkaline solution, thereby reducing the dissolution rate of alumina; When the effective amount of calcium oxide is below 30% (wt), the specific calcium aluminate phase in the calcined clinker is calcium aluminate and calcium hexaaluminate with poor alkali solubility; When the effective amount of calcium oxide is between 40% (wt) and 50% (wt), the calcium aluminate component in the calcined clinker is dodecalcium heptaaluminate with good alkali solubility. At this time, the recovery rates of aluminum, sodium, and fluorine elements in the aluminum electrolyte are relatively high, reaching 93.35%, 90.21%, and 90.53%, respectively.

This article innovatively proposes a method for the recovery and utilization of aluminum electrolytes through the research on the multi element recovery and utilization of calcium calcination roasting. The method of "calcium calcination phase inversion alkali deep leaching" has been verified to be completely feasible through thermodynamic calculations. The experiment shows that the effective amount of calcium oxide added is the main influencing factor on the recovery and utilization of multiple elements in aluminum electrolyte calcium roasting. When the effective amount of calcium oxide added is below 30%(wt), the calcium aluminate phase in the calcined product is CaO · 2Al$_2$O$_3$ and CaO · 6Al$_2$O$_3$ with poor dissolution performance. When the effective amount of calcium oxide is between 40%(wt) and 50%(wt), the calcium aluminate in the calcined product is 12CaO · 7Al$_2$O$_3$ with good dissolution performance. When the effective calcium oxide addition is above 60%(wt), the main calcination products are 12CaO · 7Al$_2$O$_3$ and excess CaO. Therefore, the suitable conditions for aluminum electrolyte calcium calcination are the effective calcium oxide addition of 40%–50%(wt). Under the condition of effective calcium oxide addition of 40%–50%(wt), the recovery rates of aluminum, sodium, and fluorine in aluminum electrolyte after calcination, leaching, and separation can reach 93.35%, 90.21%, and 90.53%, respectively.

3.2.2 Effect of calcination temperature on the recovery rate of aluminum and sodium

This section investigates the effect of calcination temperature on the calcium calcination products of waste aluminum electrolyte. Determine the phase composition of the calcined products under different influencing factors using XRD, determine the content of calcium, aluminum, fluorine, sodium, and oxygen in the calcined products using XRF, and analyze the distribution characteristics of the calcined reaction products using SEM-EDS and optical microscopy.

The main component of waste aluminum electrolyte is cryolite, in addition to a portion of calcium fluoride and aluminum fluoride[11]. The principle of the roasting experiment is to use waste aluminum electrolyte and calcium oxide as raw materials, and sodium carbonate as a phase conversion additive for mixed roasting. Through roasting, the cryolite inside the electrolyte reacts with the additive and decomposes. Due to the fact that aluminum in waste aluminum electrolytes mainly exists in the form of cryolite, the crystal structure of cryolite is shown in Fig. 3-9. Through preliminary analysis of the crystal structure of cryolite, it was found that its crystal structure consists of an isolated octahedron AlF$_6^{3-}$ bound to Na$^+$ at six vertices, forming a regular octahedron with good stability, which is generally difficult to decompose[12]. When the waste aluminum electrolyte is mixed with additives for roasting, highly active calcium oxide reacts with cryolite to generate alumina. The generated alumina continues to combine with calcium oxide to form calcium aluminate, which can be dissolved through alkaline solution[13-16].

3.2 Experiment on calcification and roasting of waste aluminum electrolyte

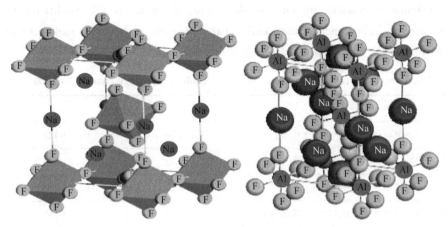

Fig. 3-9 Crystal structure of cryolite

The Gibbs free energy equations of various reactions that may occur during the calcination reaction process were calculated through FactSage, and the thermodynamic study of the calcination reaction process will provide corresponding theoretical basis for subsequent process experimental research, which is of great significance[17]. By thermodynamic analysis of the calcination process, the theoretical possibility of adding calcium oxide and sodium carbonate to treat and recover waste aluminum electrolytes is determined, and the migration law of valuable elements in waste aluminum electrolytes is analyzed.

Based on the thermodynamic calculation results of FactSage, it can be seen from Fig. 3-10 that the Gibbs free energies of reactions Eq. (3-12) – Eq. (3-18) that may occur

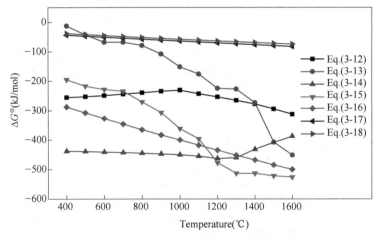

Fig. 3-10 Relationship between Gibbs free energy and temperature in calcination reaction (Table 3-4) of calcification

during the calcination process are far less than zero, which meets the conditions for reaction occurrence. This indicates that adding calcium oxide and sodium carbonate to treat and recycle waste aluminum electrolytes is thermodynamically feasible.

Table 3-4 Main reaction processes and their standard Gibbs free energy change (T = 1000 ℃)

Reactions	$\Delta G^\ominus$ (kJ/mol)	No.
$2Na_3AlF_6 + 3CaO = Al_2O_3 + 6NaF + 3CaF_2$	-230.76	Eq. (3-12)
$2Na_3AlF_6 + 3Na_2CO_3 = Al_2O_3 + 12NaF + 3CO_2(g)$	-151.43	Eq. (3-13)
$3CaO + 2AlF_3 = Al_2O_3 + 3CaF_2$	-450.23	Eq. (3-14)
$3Na_2CO_3 + 2AlF_3 = Al_2O_3 + 6NaF + 3CO_2(g)$	-361.47	Eq. (3-15)
$12CaO + 7Al_2O_3 = 12CaO \cdot 7Al_2O_3$	-400.47	Eq. (3-16)
$CaO + 6Al_2O_3 = CaO \cdot 6Al_2O_3$	-63.84	Eq. (3-17)
$CaO + 2Al_2O_3 = CaO \cdot 2Al_2O_3$	-57.93	Eq. (3-18)

Experimental conditions: The addition amount of calcium oxide is 80% (wt) of the waste aluminum electrolyte, the addition amount of sodium carbonate is 10% (wt) of the total mass, and the roasting time is 1 h. The roasting temperatures are 900 ℃, 950 ℃, 1000 ℃, 1050 ℃, and 1100 ℃, respectively.

Fig. 3-11 and Fig. 3-12 show the XRD analysis of the products obtained from the

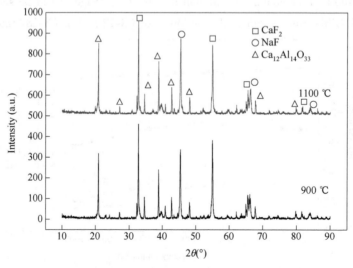

Fig. 3-11 XRD patterns of products at different calcination temperatures
(900 ℃, 1000 ℃)

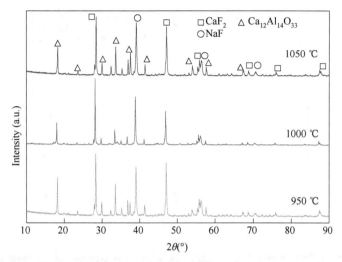

Fig. 3-12 XRD patterns of products at different calcination temperatures
(950 ℃, 1000 ℃, 1050 ℃)

calcination of waste aluminum electrolyte at different temperatures. From the figure, it can be seen that the phase composition of the calcined products obtained at different calcination temperatures is basically consistent, mainly CaF_2, NaF, and $12CaO \cdot 7Al_2O_3$. From this, it can be seen that the calcination temperature has little effect on the composition of the calcium calcined products of waste aluminum electrolyte.

The sintered products at different calcination temperatures were stirred and dissolved in a 5%(wt) NaOH solution at a liquid-solid ratio of 10 : 1 and a dissolution temperature of 90 ℃ for 90 min. The effect of different calcination temperatures on the phase composition of the leached products was analyzed.

The effect of different roasting temperatures on the phase composition of the leaching products is shown in Fig. 3-13. From the figure, it can be seen that the phase composition of the alkaline leaching products after calcination between 900 ℃ and 1100 ℃ is a single CaF_2. At 900 ℃, the characteristic peak intensity of the leaching product CaF_2 phase is significantly lower than other temperatures. It is speculated that the aluminum and sodium recovery rates under 900 ℃ conditions are lower than other temperature conditions, and the purity of the obtained calcium fluoride product is lower than other temperature conditions.

The effects of different calcination temperatures on the recovery rate of aluminum and sodium are shown in Fig. 3-14 under the conditions of dissolution temperature of 90 ℃, dissolution time of 90 min, NaOH concentration of 5%(wt), and liquid-solid ratio of 10 : 1.

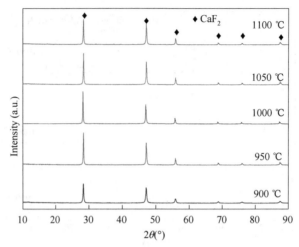

Fig. 3-13 Effect of roasting temperature on the composition of leaching products

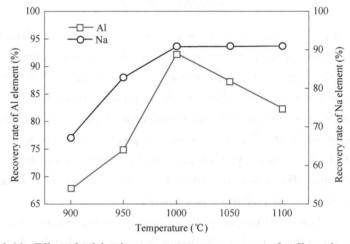

Fig. 3-14 Effect of calcination temperature on recovery of sodium aluminum

From the graph, it can be seen that the recovery rate of aluminum shows a trend of first increasing and then decreasing. When the roasting temperature is low, the recovery rate of aluminum is lower. This may be because the low temperature is not enough to completely react between cryolite and calcium oxide, resulting in some aluminum being less easily leached by alkaline solution. It may also be because the reaction between cryolite and calcium oxide produces a calcium aluminate phase with poor alkali solubility. Within the temperature range of 900–1000 ℃, the aluminum recovery rate of the mixture significantly increased, from 67.69% to 92.31%. This may be due to the fact that with the increase of temperature, thermochemical reactions are more easily carried out, generating a phase that is easily soluble in alkaline solution. Within the temperature range of 1000-1100 ℃, the aluminum dissolution rate decreases, which may be due to the increase in temperature causing the generated calcium aluminate phase (dodecalcium heptaaluminate) crystal grains to become larger, resulting in a decrease in crystal defects and an increase in the stability of dodecalcium heptaaluminate crystals[18]. This slightly reduces the ability of NaOH and H_2O molecules in the solution to penetrate into the dodecalcium heptaaluminate lattice and undergo reactions, to some extent, it reduces the leaching performance of the clinker.

3.2.3 Effect of sodium carbonate addition on aluminum and sodium recovery rates

Experimental conditions: Calcination temperature 1000 ℃, calcination time 1 h, addition of 80% (wt) calcium oxide to waste aluminum electrolyte, and addition of 0%, 5%, 10%, 15%, and 20% (wt) sodium carbonate to the total mass of the material.

Fig. 3-15 and Fig. 3-16 show the XRD analysis of the products obtained from the calcination of waste aluminum electrolytes with different amounts of sodium carbonate added. From the figure, it can be seen that the phase composition of the calcined products obtained at different amounts of sodium carbonate is basically consistent, mainly CaF_2, NaF, and $12CaO \cdot 7Al_2O_3$. From this, it can be seen that the amount of sodium carbonate added has little effect on the phase composition of the calcined products of waste aluminum electrolyte.

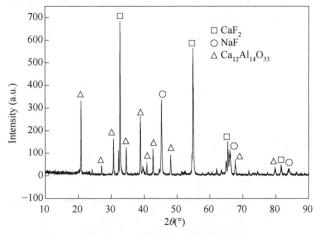

Fig. 3-15 Effect of sodium carbonate addition (0%) on composition of calcined products

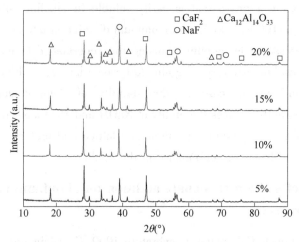

Fig. 3-16 Effect of sodium carbonate addition
on composition of calcined products (wt)

The calcined products were stirred and dissolved in a 5% (wt) NaOH solution at a liquid-solid ratio of 10 : 1 and a dissolution temperature of 90 ℃ for 90 min. The effects of different amounts of sodium carbonate added on the phase composition of the leached products were analyzed.

The effect of different amounts of sodium carbonate added on the phase composition of the leaching product is shown in Fig. 3-17 and Fig. 3-18. From the figure, it can be seen that when the amount of sodium carbonate added is between 0-20%, the crystal phase composition of the filter residue obtained from the alkaline leaching of the waste

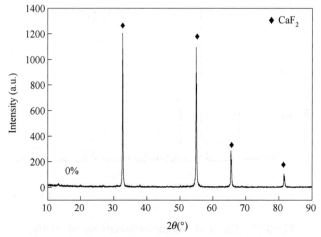

Fig. 3-17 Effect of sodium carbonate addition on
the composition of alkali leaching products

aluminum electrolyte after calcination is a single CaF_2 phase[19]. The addition of sodium carbonate has little effect on the phase composition of the leaching product.

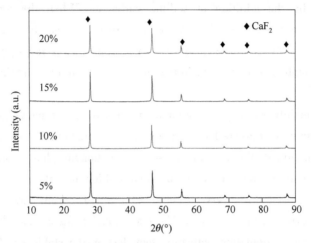

Fig. 3-18 Effect of sodium carbonate addition on
the composition of alkali leaching products(wt)

The effect of sodium carbonate addition on the recovery rate of aluminum sodium is shown in Fig. 3-19 under the conditions of dissolution temperature of 90 ℃, dissolution time of 90 min, NaOH concentration of 5%(wt), and liquid-solid ratio of 10 ∶ 1.

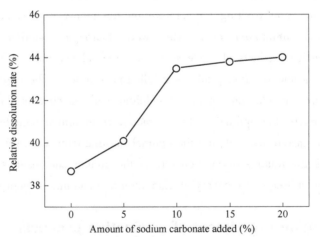

Fig. 3-19 Effect of sodium carbonate addition
on clinker dissolution rate

From the figure, it can be seen that as the amount of sodium carbonate added increases, the relative dissolution first increases and then tends to stabilize. When the amount of sodium carbonate added increased from 0% to 10%, the relative dissolution rate increased from 38.7% to 43.5%, and the relative dissolution rate increased by 4.8%. This may be due to the promotion of the formation of seven aluminum twelve calcium phases by the addition of sodium carbonate. When the amount of sodium carbonate added increases from 10% to 20%, the relative dissolution rate increases from 43.5% to 44%, and the increase in relative dissolution rate is not significant. This may be due to the complete reaction transformation of the aluminum containing phase in the raw material[20].

This chapter mainly studies the calcination process of extracting multiple elements from waste aluminum electrolyte by calcination and calcination. The effects of the amount of calcium oxide added, the amount of sodium carbonate phase conversion additive added, and calcination temperature on the composition of calcined products, alkaline leaching products, and the recovery rate of aluminum and sodium after treatment of waste aluminum electrolyte were investigated. The main conclusions are as follows:

(1) Calcium oxide calcination activation can destroy the stable cryolite phase in the waste aluminum electrolyte, generate calcium aluminate that can dissolve in alkaline solution, and fully activate the waste aluminum electrolyte. The optimal process conditions are: 80% (wt) calcium oxide addition, 10% (wt) sodium carbonate phase conversion additive addition, and calcination temperature of 1000 ℃.

(2) During the roasting experiment, the effects of calcium oxide addition, sodium carbonate addition, and roasting temperature on the recovery rates of aluminum and sodium in waste aluminum electrolytes were thoroughly studied, revealing the calcification roasting products of waste aluminum electrolytes under different conditions and their leaching and reaction products in alkaline solutions. The experimental results show that the optimal addition amount of calcium oxide is 80% of the mass of waste aluminum electrolyte, the optimal addition amount of sodium carbonate is 10% of the total mass of the mixed material, and the optimal roasting temperature is 1000 ℃. Under these conditions, the relative dissolution rate of the clinker can reach 43.5%, which is conducive to the subsequent recovery of aluminum and sodium elements.

3.3 Dissolution experiment of calcination products

Calcium aluminate in calcined clinker is easily soluble in alkaline solution, achieving the extraction of alumina resources. Furthermore, the harmful substance fluoride will be solidified into the form of calcium fluoride, achieving the removal of toxicity (the reaction formula is shown below).

$$12CaO \cdot 7Al_2O_3 + 14NaOH + 33H_2O = 14NaAl(OH)_4 + 12Ca(OH)_2 \quad (3-19)$$
$$Ca(OH)_2 + 2NaF = CaF_2 + 2NaOH \quad (3-20)$$
$$CaO \cdot 6Al_2O_3 + 12NaOH + 19H_2O = 12NaAl(OH)_4 + Ca(OH)_2 \quad (3-21)$$
$$CaO \cdot 2Al_2O_3 + 4NaOH + 7H_2O = 4NaAl(OH)_4 + Ca(OH)_2 \quad (3-22)$$

From the above reaction mechanism, it can be seen that the main components of the filtrate are sodium aluminate and sodium hydroxide, while the main component in the insoluble matter is calcium fluoride. Therefore, after solid-liquid separation, aluminum and sodium can be recovered.

Specific experimental methods: Weigh 10 g of roasted clinker under the optimal roasting process conditions and grind for 30 min. Heat the dilute alkali solution to a certain temperature, and then add the clinker in the stirring process. Based on the leaching experiments of clinker under three different process conditions: Sodium hydroxide concentration, leaching temperature, and leaching time, the recovery rates of aluminum and sodium are measured to determine the optimal alkaline leaching conditions.

3.3.1 Effect of dissolution temperature

Experimental conditions: Sodium hydroxide concentration of 5% (wt), dissolution time of 90 min, liquid-solid ratio of 10 : 1, dissolution temperatures of 30 ℃, 50 ℃, 70 ℃, 90 ℃, and 100 ℃, respectively.

As shown in Fig. 3-20, the relationship between dissolution temperature and aluminum sodium recovery rate is described. As the reaction temperature increases, the aluminum oxide dissolution rate of calcium aluminate in the calcined product gradually increases. It can be seen that the temperature has a significant impact on the aluminum oxide dissolution rate. When the temperature increases from 30 ℃ to 90 ℃, the recovery rate of aluminum sodium increases by 16.14% and 6.8%, respectively. This may be due to the acceleration of reaction molecular motion rate and mass transfer between liquid and solid phases as the temperature increases; After 90 ℃, the recovery rate of aluminum and sodium does not increase significantly. Moreover, when the temperature rises to 100 ℃, the reaction system evaporates quickly, loses more water, and the viscosity increases, affecting the molecular motion rate. Therefore, although the temperature increases, the change in the recovery rate of aluminum and sodium is not significant.

Therefore, in order to achieve better dissolution effect and avoid excessive energy consumption, 90 ℃ was selected as the optimal dissolution temperature in the experiment.

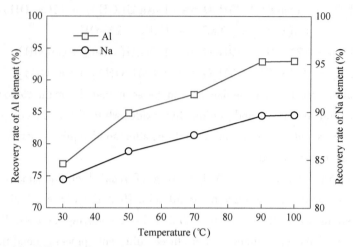

Fig. 3-20 Effect of dissolution temperature on recovery of sodium aluminum

3.3.2 Effect of dissolution time

Experimental conditions: Sodium hydroxide concentration of 5% (wt), dissolution temperature of 90 ℃, liquid-solid ratio of 10 : 1, dissolution time of 30 min, 50 min, 70 min, and 90 min, respectively.

As shown in Fig. 3-21, the relationship between dissolution time and aluminum sodium recovery rate is described. From the graph, it can be seen that with the increase of dissolution time, the recovery rate of aluminum and sodium shows an increasing trend, with a significant increase in 0-30 min. This may be due to the sufficient amount of calcined products and sodium hydroxide raw materials in the early stage of the

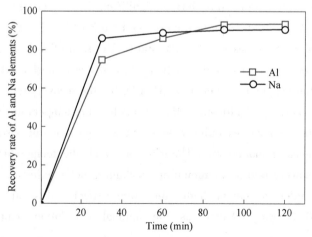

Fig. 3-21 Effect of dissolution time on recovery of sodium aluminum

experiment, resulting in a fast reaction rate of the system, so the dissolution rate of aluminum oxide increases faster. When the dissolution time increases from 30 min to 90 min, the recovery rate of aluminum sodium still increases, but the rate of increase is slow. The dissolution time continues to extend, indicating that the recovery rate of aluminum sodium does not change significantly, possibly due to the complete reaction between dodecalcium heptaaluminate and sodium hydroxide in the calcined product. Therefore, in order to maintain dissolution efficiency and avoid excessive cost consumption, 90 min was selected as the optimal dissolution time in the experiment.

3.3.3 Effect of alkali concentration

Experimental conditions: The dissolution temperature is 90 ℃, the dissolution time is 90 min, the liquid-solid ratio is 10 : 1, and the concentration of sodium hydroxide is 0%, 5%, 10%, 15%, and 20% (wt), respectively.

As shown in Fig. 3-22, the relationship between sodium hydroxide concentration and aluminum sodium recovery rate is described. From the figure, it can be seen that as the concentration of sodium hydroxide increases, the dissolution effect of alumina gradually becomes apparent. When the concentration of sodium hydroxide increases from 1% to 3% (wt), the recovery rate of aluminum sodium significantly increases. This may be because the amount of sodium hydroxide is insufficient at the beginning of the reaction. As the concentration of sodium hydroxide increases, the amount of dodecalcium heptaaluminate consumed in the reaction gradually increases, indicating an increase in the dissolution rate of alumina; When the concentration of sodium hydroxide increases

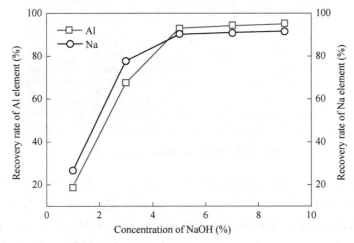

Fig. 3-22 Effect of NaOH concentration on recovery rate of aluminum and sodium

from 3% to 5% (wt), the recovery rate of aluminum sodium still increases, but the increase rate slows down compared to before; When the concentration of sodium hydroxide is greater than 5% (wt), the recovery rate of aluminum sodium does not change significantly. This may be because the soluble calcium aluminate in the calcined product reacts completely with sodium hydroxide in the solution, and increasing the concentration of the alkaline solution has little effect on the recovery rate of aluminum sodium. Therefore, 5% (wt) is the optimal concentration for the alkali solution experiment.

3.3.4 Analysis of microscopic morphology of dissolved products

The microstructure of the dissolved products was analyzed under the roasting conditions of 80% (wt) calcium oxide, 10% (wt) sodium carbonate phase conversion additive, 1000 ℃ roasting temperature, and 90 ℃ dissolution temperature, 5% sodium hydroxide concentration, and 90 min dissolution time. Fig. 3-23 is a light microscope photo of the dissolved product, which shows that the particle size distribution of the product calcium fluoride is relatively uniform, and the agglomeration phenomenon may be due to insufficient dispersion in ethanol.

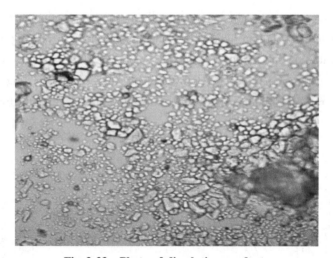

Fig. 3-23 Photo of dissolution products

Further analysis was conducted using SEM-EDS to investigate the microstructure and composition of the dissolved product calcium fluoride. Fig. 3-24 and Fig. 3-25 show the SEM-EDS results of the alkaline leaching residue. Fig. 3-24 (a) – (d) show that the samples were dispersed on silicon wafers and magnified 1000, 2000, 3000 and 5000 times, respectively. It can be seen that the dissolution product particles are evenly dispersed. Fig. 3-24(e) and (f) show the sample being coated on conductive adhesive and magnified by 2000-3000 times.

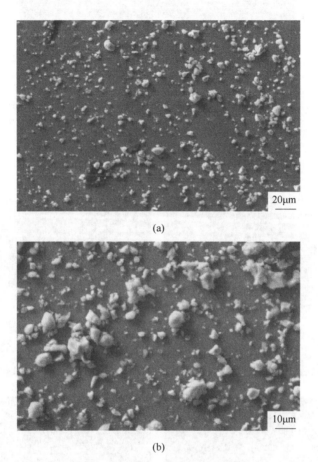

Fig. 3-24   Electron micrograph of dissolution product
(a) Scale of 20 μm; (b) Scale of 10 μm

From Fig. 3-25, it can be seen that the surface scanning results of the alkali leaching slag show that the main elements in the slag are Ca and F, and there are also small amounts of Na and Al remaining, indicating that the calcium aluminate in the calcined product is basically completely reacted and the alkali dissolution effect is good.

Fig. 3-26 show SEM-EDS images of alkali leaching residue. According to the EDS spectra of points 1 and 2, it can be seen that the main phase in the alkali leaching residue is calcium fluoride, while there are also traces of sodium and aluminum remaining.

This chapter mainly studies the alkaline leaching process of extracting multiple elements from waste aluminum electrolyte by calcination and calcination, and investigates

Fig. 3-25 Surface scan photos of dissolution products

the effects of dissolution temperature, dissolution time, and sodium hydroxide concentration on the recovery rate of aluminum and sodium in waste aluminum electrolyte. The main conclusions are as follows:

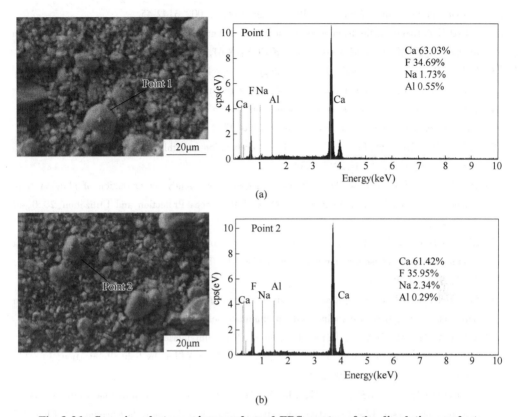

Fig. 3-26 Scanning electron micrographs and EDS spectra of the dissolution products

(1) In the alkaline leaching process, the most important factor affecting the recovery rate of aluminum and sodium is the concentration of sodium hydroxide. The influence of dissolution temperature and time on the recovery rate of aluminum and sodium is relatively small.

(2) The most suitable alkaline leaching conditions are: dissolution temperature 90 ℃, sodium hydroxide concentration 5%, dissolution time 90 min, and the recovery rates of aluminum and sodium can reach 93.35% and 90.21%, respectively.

(3) SEM-EDS analysis of the dissolved products revealed that the residual Na content in the alkaline leaching residue was 1.7%–2.3%, and the residual Al content was 0.3%–0.6%, indicating a good alkaline leaching effect; The main component of alkaline leaching residue is CaF_2, which can be used as a raw material for the production of fluoride salts.

References

[1] WANG P W, LIU F Q, TIAN C Q. Hazards and influencing factors of carbon anodizing and slag

cracking in aluminum electrolysis cell[J]. Light Metals,2002,3:42-45.

[2] ZHAO Z Y. Exploring the technology of energy saving and emission reduction in primary aluminum production[J]. World Nonferrous Metals,2019,8:28-31.

[3] LI H F. Research on the application technology of electrolyte powder instead of cryolite in the firing start-up of aluminum reduction cell[D]. Northeastern University,2011.

[4] WANG J P,YE J M,LIN L F,et al. Research on lithium extraction process from primary aluminum waste residue[J]. Henan Chemical Industry,2020,37(1):36-39.

[5] WANG W, CHEN W, LI Y, et al. Study on preparation of lithium carbonate from lithium-rich electrolyte[J]. Light Metals,2019:923-927.

[6] HUAN S X, WANG Y W, DI Y Z, et al. Experimental study on extraction of alumina from secondary aluminum ash calcination[J]. Mineral Resources Protection and Utilization, 2020, 40(3):34-39.

[7] ZHANG D, GU L, LIU E K, et al. High-temperature sintering non-steady state sodium-containing calcium aluminate phase transformation mechanism[J]. The Chinese Journal of Nonferrous Metals, 2019,29(8):1740-1748.

[8] SUN H L,TU G F,BI S W,et al. Synthesis of seven-aluminum twelve calcium and its dissolution performance in high-carbon sodium sodium aluminate solution[J]. Light Metal,2007,11:17-19.

[9] TODA Y, MIYAKAWA M, HAYASHI K, et al. Thin film fabrication of nano-porous 12CaO · 7Al$_2$O$_3$ crystal and its conversion into transparent conductive films by light illumination[J]. Thin Solid Films,2003,445(2):309-312.

[10] CHATTERJEE A, NISHIOKA M, MIZUKAMI F. A periodic first principle study to design microporous crystal 12MO · 7Al$_2$O$_3$ for selective and active O$^-$ radicals encaging[J]. Chemical Physics Letters,2004,390(4):335-339.

[11] PANAITESCU A,MORARU A,PANAITESCU I. Research on the instabilities in the aluminum electrolysis cell[J]. Light Metals,2003:359-366.

[12] HAARBERG G M, OSEN K S, THONSTAD J, et al. Measurement of electronic conduction in cryolite alumina melts and estimation of its effect on current efficiency [J]. Metallurgical Transactions B,1993,24(5):729-735.

[13] KVANDE H,MOXNES B P,SKAAR J,et al. Pseudo resistance curves for aluminum cell control-alumina dissolution and cell dynamics[J]. Light Metals,2016:760-766.

[14] HOMSI P, PEYNEAU J M, REVERDY M. Overview of process control in reduction cells and potlines[J]. Light Metals,2016:739-746.

[15] GRJOTHEIM K,MATIASOVSKY K,MALINOVSKY M. Influence of NaCl and LiF on the aluminium electrolyte[J]. Electrochimica Acta,1970,15(2):259-269.

[16] ROLSETH S. Low temperature aluminium electrolysis in a high density elektrolyte[J]. Aluminium, 2005,81(5):448-450.

[17] KVANDETECHN H. Bath chemistry and aluminum cell performance—Facts,fictions,and doubts [J]. JOM,1994,46(11):22-28.

[18] GAO B L,ZHAO Q,JIANG Q W,et al. Phase equilibrium on aluminum electrolyte and determination

of cryolite ratio[J]. The Chinese Journal of Nonferrous Metals,2002,12(5):1055-1059.
[19] YURKOV V,MANN V,PISKAZHOVA T,et al. Dynamic control of the cryolite ratio and the bath temperature of aluminium reduction cell[J]. Light Metals,2002,131:383-388.
[20] CHEN P,HOU P P,ZHAI J H,et al. A novel method for the comprehensive utilization of iron and titanium resources from a refractory ore[J]. Separation and Purification Technology,2019,226: 1-7.

Chapter 4 Outlook

After 65 years of development, especially the rapid development in the past 30 years, China's aluminum industry has achieved remarkable results. The production of alumina, primary aluminum, carbon for aluminum, and recycled aluminum accounts for more than half of the world's total production, and the main production technology indicators have reached the world's advanced level. On the other hand, the rapidly developing aluminum industry has brought about a series of problems such as severe overcapacity and unreasonable layout, resource depletion, energy shortage, and environmental pollution.

The continuous innovation capacity of China's aluminum industry is insufficient. Due to resource and energy issues, the core competitiveness of products is insufficient. Despite overcapacity, aluminum consumption is showing a downward trend, and environmental pollution is severe. However, there is a lack of support from key technologies for governance. This is the main structural obstacle that currently exists in China's aluminum industry.

China's aluminum industry is at an important historical turning point. Against the backdrop of vigorously promoting supply side reform and advocating ecological civilization construction, China's aluminum industry can only fundamentally coordinate the relationship between resources, energy, and environment, take the path of energy conservation, low consumption, high-quality, and environmental protection, enhance its core competitiveness, and promote the high-quality development of the aluminum industry. Simultaneously following the trend of economic globalization, promoting the optimized flow and combination of production factors, and creating a global industrial chain and market environment that is mutually collaborative, complementary, and deeply integrated, can achieve long-term and stable sustainable development.

Based on a comprehensive analysis of the current situation and existing problems of China's aluminum industry, this report mainly elaborates on the direction and strategic recommendations for the sustainable development of China's aluminum industry from three aspects: bauxite resources, energy, core competitiveness, and environment.

4.1 Resource and energy development strategy of China's aluminum industry

Under the guidance of the the Belt and Road Initiative, we will focus on gathering overseas high-quality bauxite resources to provide China's alumina industry with high-quality and stable bauxite raw materials. The key development areas of foreign bauxite are West Africa (Guinea, Ghana and other countries), Australia and Southeast Asia (Indonesia, Malaysia, Laos, Cambodia and other countries). On the other hand, for the complex composition of diaspore ore in China, relevant technologies have been developed to remove harmful impurities and minerals, achieving efficient and low consumption utilization. China's aluminum electrolysis production capacity is selectively transferred to major hydropower provinces such as Yunnan and Sichuan, or coal resource provinces such as Inner Mongolia. Deep cooperation is implemented with domestic hydropower or thermal power enterprises to form green and low-priced power supply, improve energy cost competitiveness, and seek opportunities to develop in regions with particularly abundant energy (including hydropower) such as Central and West Africa, the Middle East, and Latin America. China's aluminum industry is located in politically stable, resource and energy rich countries and regions around the world. It can adopt various flexible methods such as holding shares, holding shares, and participating in operations to engage in multi-dimensional cooperation with foreign aluminum industry enterprises, achieving mutual integration, mutual benefit, and win-win situation in the domestic and foreign aluminum industry. This is an important way for China's aluminum industry to achieve structural layout adjustment and sustainable development.

4.2 Strategy for improving core competitiveness of China's aluminum industry

The production capacity scale of China's aluminum industry should be matched with economic development and resource and energy conditions, with the main goal of meeting basic domestic demand. Based on the development trend of domestic and international demand for aluminum products and the actual situation of domestic production capacity, it is recommended that the country strictly control the total production capacity and output of industries such as alumina, aluminum electrolysis, and aluminum carbon in the aluminum industry. In recent years, the production of raw aluminum, alumina, and aluminum carbon products should be controlled at around 36 Mt, 72 Mt, and 20 Mt respectively. Under the condition of total quantity control, China's alumina industry mainly utilizes local bauxite in mainland provinces, while alumina factories in coastal areas use imported mines. The newly added alumina production capacity along the coast replaces the outdated production capacity eliminated in the mainland. At the appropriate time, some alumina production capacity can be transferred to regions rich in bauxite such as West Africa, Southeast Asia, and South America. Develop a quality development strategy for high-quality alumina products and strictly implement the quality standards for sandy alumina. Develop and promote technologies to improve cycle efficiency and deep energy conservation and emission reduction, eliminate outdated production capacity based on energy consumption, production efficiency, and product quality standards, and develop technologies that can economically utilize complex diaspore minerals to enhance competitiveness. Under the condition of total quantity control, China's aluminum electrolysis industry implements the optimization of production capacity allocation with the use of green and low-priced energy as the core, transferring some production capacity to areas with green energy and cheap electricity prices. Further optimize the physical field simulation and intelligent design technology of large aluminum electrolysis cells, develop aluminum electrolysis technologies that improve current efficiency and reduce material consumption, and eliminate outdated production capacity based on energy consumption and production efficiency standards. Vigorously improve the intelligent control level of aluminum electrolysis, achieve deep energy conservation, consumption reduction, and emission reduction. China's aluminum carbon industry should make good use of the complex composition of petroleum coke raw materials, focus on developing and applying production technologies of high-quality carbon anodes and graphitized carbon cathodes, and further achieve the goals of energy conservation, quality improvement, and clean production. Vigorously develop China's recycled aluminum industry and further form a large-scale industry. Strengthen national top-level design, establish a sound waste aluminum recycling management system, accelerate the realization of clean production, energy conservation and emission reduction of recycled aluminum, and improve core competitiveness. Develop high-value aluminum dross disposal and deep flue gas purification technologies to achieve green development of China's recycled aluminum industry. Vigorously develop automation, informatization, and intelligent technologies in various industries of China's aluminum industry, accelerate the construction of China's aluminum industry intelligent manufacturing system, especially vigorously develop sensitive primary components and intelligent system software, and quickly establish intelligent demonstration enterprises to significantly improve production efficiency, achieve energy conservation, emission reduction, and clean production, and improve safety production, stable control, and green environmental protection levels.

4.3　Environmental development strategy of China's aluminum industry

The treatment of red mud includes three major aspects: Strict anti-seepage and dam collapse prevention in the red mud yard, implementation of grass and tree planting and reclamation after the closure of the reservoir, and utilization of red mud resources. The main direction for the safety and environmental protection disposal of red mud yards is to improve the safety and reliability of dry red mud piles, strengthen online monitoring and safety control around the yard, develop bio greening of high salt alkali red mud yards, and reclamation or reuse technologies after the red mud yard is closed. The direction of resource utilization of red mud is a technological route that can be economically consumed on a large scale. For example, it is used in the fields of steel, building materials, and environmental protection. The products produced can utilize valuable elements in red mud and achieve the solidification or removal of harmful elements. Vigorously developing safe disposal and high-value resource utilization technologies for solid waste from aluminum electrolysis, including aluminum dross, waste cathode carbon blocks, waste side lining, waste refractory materials, and carbon slag, must be classified and treated according to the characteristics of different waste to achieve efficient separation of components with different properties and achieve high-value utilization of each component. Vigorously develop solid waste resource utilization technologies generated by the recycled aluminum industry, especially to accurately separate and utilize valuable components such as aluminum, sodium, fluorine, carbon, and heavy non-ferrous metals in these waste. Vigorously develop technologies for the treatment of waste gases containing fluorine and sulfur in aluminum electrolysis, carbon for aluminum use, and the recycled aluminum industry, especially to solve the problem of unorganized emissions in the production process. Develop intelligent control technology for aluminum electrolysis cells to reduce the coefficient of effect and reduce greenhouse gas emissions from perfluorocarbons. Develop efficient dioxin purification technologies to remove sulfur, asphalt, and regenerated aluminum production emissions from flue gas, and achieve standard emissions of all exhaust gases from China's aluminum industry.